Mainstreaming Science and Mathematics

Mainstreaming Science and Mathematics:
SPECIAL IDEAS AND ACTIVITIES FOR THE WHOLE CLASS

Charles R. Coble
East Carolina University

Paul B. Hounshell
University of North Carolina
at Chapel Hill

Anne H. Adams
Duke University

Goodyear Publishing Company, Inc.   Santa Monica, California

Library of Congress Cataloging in Publication Data

Coble, Charles R
    Mainstreaming science and mathematics.

    Companion vol. to the authors' Mainstreaming language
arts and social studies.
    1. Science—Study and teaching (Elementary)
2. Mathematics—Study and teaching (Elementary)
I. Hounshell, Paul B., joint author. II. Adams,
Anne H., joint author. III. Title.
LB1585.C57   372.3'5'044   76-50042
ISBN 0-87620-599-6

Current printing (last digit):
10 9 8 7 6 5 4 3 2 1

ISBN: 0-87620-599-6 (P)
       0-87620-592-9 (C)
       Y-5996-7 (P)
       Y-5929-8 (C)

Book production by Ken Burke & Associates

Project editor: Judith Fillmore
Text and cover designer: Christy Butterfield
Illustrator: Nancy Freeman
Compositor: CBM Type

Printed in the United States of America

To Ann, Diana, Lois, and Mac

Preface

Before mainstreaming, elementary teachers had in their classes students displaying a wide range of abilities. After mainstreaming, in most classes, the range became wider. This book is written for elementary teachers who want to improve their strategies for working with a wide variety of students' abilities and interests. This variety now includes students formerly assigned all day to special-education classes.

Dedicated teachers are concerned with how best to minimize lines of division between regular classroom students and special-education students. They request ideas that help bring these students together and avoid situations that make the differences more marked than is absolutely necessary. Just as importantly, they want a curriculum that moves *all* students forward during the year—in academics, in accepting responsibility, in working with others, and in understanding their own actions and thoughts.

This book marks a new day in instruction suggestions for the following reasons:

No compartmentalization—No longer is it realistic to mark one set of ideas "Special Education" and another set "Other Fifth-Grade Students." Therefore, this book is *not* designed just for exceptional children and it is *not* designed only for other elementary students. It is intended for *all* students in any elementary class, and it is intended to be taught by the professional teacher.

No author-determined or test-determined grade level—Although many of the ideas in this book can be used with junior-high students, the thrust is elementary level. Children frequently surprise teachers. In researching activities for this book, we found students who easily completed lessons when all prior indications were that they would not be able to do them. On the other hand, some of the apparently easier activities were difficult for students who where supposed to "sail through" them with ease. *No one can accurately predict which lessons students can and cannot do well.* At the Duke University Reading Center, where many of these ideas were used with students ranging from retarded to gifted, the major conclusion reached was that one cannot teach students unless one exposes them to the content. Similar results were found in the reactions of teachers and students at the University of North Carolina at Chapel Hill and at East Carolina University. Therefore, this book has no grade limits; that determination is not nearly as important as what takes place between teacher and student.

A wide, rather than restricted, learning base—Obviously, every student will not complete every activity in this book; however, the philosophy in this book is to present numerous teaching ideas. An individual student cannot and should not be pressured to learn too much in a short period of time, but the opposite situation places the educator in a dilemma. Slow-paced, highly repetitive lessons are usually deadly. We have chosen many ideas for teachers to select from, rather than a few ideas for teachers to try to stretch. We have found that most teachers do not overload their students. Consider this book a treasure chest of teaching ideas, and pace the activities according to professional judgment.

Print and paper as major tools—It is quite popular in education today to ask teachers to use a variety of materials and teaching strategies—and most teachers do. In this book we have not emphasized the use of film, tapes, records, transparencies, etc. because we find most teachers are using these devices. We enthusiastically endorse using audiovisual materials, and there is an unwritten assumption that audiovisual materials will be used in the classroom. This book deals directly with activities for students to do using printed materials, paper, pencil, and chalkboard. We have attempted to minimize educational jargon in the book and to direct the teacher's attention to purpose, material, and possible lesson procedure.

A joy in teaching and learning—Although the book is well stocked with ideas, the real purpose is to transmit exciting learning signals to students. Only by creating situations in which different kinds of learning can take place can today's teacher hope to make great strides in instruction. Real life itself is filled with numerous series of short-term and long-term activities, with variety and with a multitude of "things." Children labeled as emotionally disturbed or perceptually handicapped should not be denied the excitement of creative and basic school-type activities; rather, all students should be given numerous content-area themes to investigate.

Specific instructional design—We find most teachers want teaching suggestions in a succinct fashion, and simply do not have time to wade through excess words. The book format is deliberately to the point, and loaded with practical ideas that apply to exceptional children working in classrooms with other children. In essence, this book is a compilation of activities emphasizing the content through which processes are oriented. Because of the variety of information, materials, individual and group techniques, etc., the threat of boredom is minimized, and the keynote is placed where it should be: transmitting knowledge, developing skills, and initiating interests. Although we realize that there are other kinds of teaching materials, we place an educational value on items that historically have served well in the classroom: paper, books, other printed materials, pencils, chalk, and chalkboard.

Time-span objectives—Individual student objectives for each theme are identified on a daily basis, rather than by week, month, or semester; the theme objectives are presented on a weekly basis. The teacher should decide how much of the week, as well as how much of the day, to devote to various aspects of the theme. Research undertaken in preparation of this book revealed that, contrary to some practices, educators do not have unlimited time to teach exceptional children. The point is not how many times an idea is repeated, or how many lessons are spent on getting ideas across to students, but rather how well a student wants to do and can do many different kinds of activities. Instead of negative planning, or not progressing until mastery is achieved, this book advocates a positive span of objectives within the realistic framework of the academic week. The theme builds during the week in an effort to accommodate the various abilities of students in the classroom. Some will zero in on the concept on Monday; others will grasp it on Friday; still others will need additional time. The decision is then left to the teacher as to whether to continue the theme for additional time or to initiate a new theme, and this can only be the teacher's decision.

We wish to thank Christy Butterfield for the book design and Nancy Freeman for the illustrations used in the book. We also wish to thank Kathleen Burroughs, Karen Collier, and Jean Taylor for typing the manuscript.

The companion volume to this book is *Mainstreaming Language Arts and Social Studies*.

Charles R. Coble
Paul B. Hounshell
Anne H. Adams

Contents

Part Two   Mathematics

Mainstreaming Science and Mathematics

Science

Science

1 Describing Colors, Sizes, and Shapes of Objects

2 Describing Textures, Odors, and Sounds of Objects

3 Investigating the Structure of Matter

4 Observing Changing States of Matter

5 Observing Changes in Matter and Molecular Motion

6 Learning about Friction

7 Transferring Heat Energy

8 Learning about Light

9 Bending Light

10 Observing Some Characteristics of Color

11 Learning about Electricity

12 Making Circuits, Conductors, and Electromagnets

13 Describing Magnets and Magnetism

14 Learning about Sound

15 Investigating Gravity

16 Studying the Solar System

17 Describing Some Effects of the Earth's Motions

18 Describing the Moon and Its Motions

19
Investigating Heating and Cooling of the Earth

20
Investigating Water in the Air

21
Collecting and Recording Weather Information

22
Identifying Factors That Weather Rock

23
Investigating Erosion

24
Identifying Green Plants

25
Studying How Seed Plants Reproduce

26
Germinating Seeds

27
Investigating Life Needs of Green Plants

28
Identifying Nongreen Plants (Fungi)

29
Identifying Animals without Backbones

30
Identifying Animals with Backbones

31
Observing Animal Needs and Reproduction

32
Identifying Survival Adaptations

33
Studying the Interdependency of Plants and Animals

34
Studying Communities of Living Things

35
Studying the Human Body Systems: Part I

36
Studying the Human Body Systems: Part II

Describing Colors, Sizes, and Shapes of Objects

General Overview The activities of the first week may serve a diagnostic function. Many students may already know the basic colors and shapes and be able to describe objects according to relative size; those who do not should be given additional learning opportunities.

Learning Objectives Students will describe objects according to their color, size, and shape. They will apply their knowledge by identifying objects related to protection and safety in the environment.

MON

Objective Describing objects according to color.

Teacher Preparation Make a display of several solid-colored objects familiar to the students. Label several sheets of colored construction paper with the name of its color.

Activities Students observe a display of colored objects and name the colors. They then make some color labels for other objects in the room.

TUES

Objective Describing and grouping objects according to size.

Teacher Preparation Have available some blocks that are different sizes of the same shape. Encourage the use of such phrases as "larger than, but shorter than," etc.

Activities Each student is asked to arrange his or her blocks according to size differences. Then the students combine their blocks and group according to size. Students then look for other objects of different sizes.

WED

Objective Identifying and describing the shapes of objects—circles, squares, triangles, and rectangles.

Teacher Preparation Give each student a set of blocks that are circular, square, triangular, and rectangular. Use an overhead projector or chalkboard to draw the shape. The shape could also be displayed on a flannel board.

Activities Students are asked to identify the shape shown them by the teacher. They hold up the block that has the same shape. They are divided into groups to conduct a "Shape Hunt" of objects around the classroom and school that have the shapes they have learned.

What color are these objects?

Make some color labels for other objects in the room!

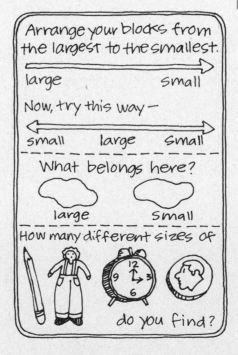

Arrange your blocks from the largest to the smallest.

large — small

Now, try this way —

small — large — small

What belongs here?

large — small

How many different sizes of do you find?

NAME THE SHAPE

SHAPE HUNT

triangles: _____

squares: _____

circles: _____

rectangles: _____

THURS

Objective Identifying and grouping objects according to color, size, and shape.

Teacher Preparation Have available a number of blocks in different sizes, shapes, and colors. Using spirit masters, make up a series of "Find the Blocks" task cards that call for the student to find blocks of one color, size, and shape. Visual clues could be provided by outlining the various sizes and shapes and by color-coding the cards with crayons or felt-tip pens.

Activities Students are given a group of blocks of different sizes, colors, and shapes. They are asked to hold up the blocks that are square and red, square and blue, triangular and green, etc. They are then asked to hold up their large, square, red block, their small, circular, blue block, etc. Students are then asked to find the blocks that correspond to the task cards.

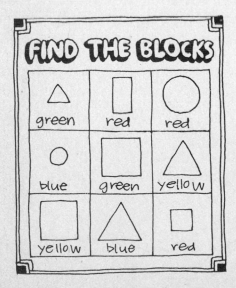

FRI

Objective Identifying colors and shapes of objects pertaining to protection and safety.

Teacher Preparation Have available colored pictures of school buses, police cars, fire trucks, ambulances, and highway signs such as stop signs, yield signs, railroad-crossing signs, speed signs, etc. An optional activity would require the use of tempera or crayons and paper.

Activities Students are shown pictures of school buses, police cars, fire trucks, and ambulances; they are asked to identify both the object and its color. They are then shown pictures of various highway signs and are asked to identify the color and shape of the signs. (The octagonal shape of the stop sign may be a shape not previously learned.) They discuss the "message" of each sign. Students then go on a "Safety Hunt," looking for colors and shapes in the vicinity of the school (stop lights, painted road lines, etc.). They draw and color some of the protection and safety-oriented objects in their environment.

5

Describing Textures, Odors, and Sounds of Objects

General Overview The activities of the second week are a continuation of the theme established during the first week: describing the properties of objects. Sensitizing students to the many different ways in which they experience their environment as well as how to communicate their experience to others is a vital experience in science instruction. The activities of this week should facilitate the development of a vocabulary that will be helpful to students in later science activities.

Learning Objectives During the second week, students will describe a variety of objects and organisms according to their texture, odor, and sound. They will also use skills gained during the first week in describing a variety of environments.

MON

Objective   Describing and comparing textures.

Teacher Preparation   Put a variety of objects (five or six) in paper lunch bags: wood blocks, wax, marbles, clay, wool and silk material (or synthetic substitutes with similar textures), nails, rocks, cotton balls, sandpaper, etc. Prepare enough bags for at least every pair of students.

Activities   Students describe to their partners how each object or substance in the bag feels, without actually naming the item. Partners then reverse roles. Then taking the objects out of the bag one at a time, the partners compare the descriptions they gave to each object. The teacher writes on the chalkboard the words used to describe the objects.

TUES

Objective   Describing and comparing odors.

Teacher Preparation   Place a variety of objects and materials inside baby-food jars. In each jar place one item such as onion, sawdust, lemon juice, a fragrant flower such as honeysuckle or gardenia, moth balls, coffee, tobacco, vinegar, cedar and/or pine needles, perfume, peppermint candy, various fruits such as apples, bananas, and oranges. Punch holes in the jar lids and wrap the jars in paper or foil so that the contents cannot be seen by the students. Place the jars at numbered stations around the room.

Activities   Students move around the room and smell the objects. They discuss their findings with the teacher, who writes on the chalkboard the words used to describe the odor at each station. The activity is concluded by unwrapping each jar to reveal the contents.

WED

Objective   Describing and comparing sounds.

Teacher Preparation   Out of view of the students, the teacher will bounce a ball, pluck a stringed instrument, drop a book, scratch the chalkboard, pour some water, whistle, or pop a balloon.

Activities   With their backs to the teacher, students listen to the sounds produced by the teacher. They identify how the sounds are made and, more importantly, describe the sound, using terms such as loud, soft, scratchy, or bubbly. There are no "correct" descriptions and everyone should be encouraged to participate.

THURS

FRI

Objective Describing the colors, shapes, sizes, textures, odors, and sounds of the school environment.

Teacher Preparation Select areas in and around the school grounds for an "experience hike" with the students: the cafeteria, the gymnasium, the health-aid room, the playground, a school bus, a wooded area, and a grassy area. Provide a list of the names of the areas to be visited, leaving space for a few words beneath each location. Divide the students into groups of three or four, with one student acting as recorder in each group. Allow enough time in each area for students to make adequate observations. Encourage them to observe more than just those things they immediately sense when entering an area.

Activities Student groups discuss their observations in each area, describing the different colors, shapes, sizes, textures, odors, and sounds that they observe. They describe the areas on paper, using words learned in previous lessons. (Complete sentences are not necessary.)

Objective Observing and inferring the properties of objects in different environments.

Teacher Preparation Seek student assistance in cutting pictures from papers and magazines that show scenes in different parts of the nation and world: mountains, oceans, beaches, deserts, jungles, cities, suburbs, polar regions, etc.

Activities Students describe the shapes seen in the pictures and make inferences about the sizes, textures, odors, and sounds they could observe in different environments.

General Overview The activities of weeks 1 and 2 introduced students to the properties by which material objects can be identified, described, and classified. The activities of this week should expand the student's concept of the three commonly existing states of matter: solids, liquids, and gases. This will establish a background from which they can learn another basic concept in science: change.

Learning Objectives Students will identify and describe characteristics of the three states of matter: solids, liquids, and gases. They will also learn that all three forms of matter take up space and have weight.

WED

Objective Identifying and describing the characteristics of gases.

Teacher Preparation Lead a brief discussion, pointing out that air is a mixture of gases. Distribute a variety of different sizes and shapes of balloons, paper bags, and plastic bags to each student.

Activities After learning that air is made up of different gases, the students blow one big breath (exhale) into each of the different objects. They observe the varied shapes and sizes of the other balloons and bags in the room. From their observations they conclude that gases have no definite size and no definite shape.

MON

Objective Identifying and describing the characteristics of solids.

Teacher Preparation Have metric rulers available for use by students.

Activities Students empty their pockets and handbags and describe the contents, using terms such as metal, wood, hard, soft, round, square, etc. They are also asked to measure the objects. Two students at a time then come to the front of the class to play "Guess the Object." With their backs to each other, one of the students holds up an object and describes it without naming it; the other student tries to guess what the object is. This is repeated around the room. The object of the lesson is sammarized by the teacher: a solid object has a definite size and shape.

TUES

Objective Identifying and describing the characteristics of liquids.

Teacher Preparation Have available several different liquids such as water, cooking oil, vinegar, syrup, and alcohol, plus many jars of different shapes and sizes.

Activities Working in groups, the students pour the different liquids into the various types of jars available. Each time they observe that the shape of the liquid conforms to the shape of the container, but the size remains the same (which they can check by pouring the liquid back into the original jar to see if the amount and shape of the liquid is the same as before it was poured out). Their observations are summarized with the conclusion that liquids have a definite size, but no definite shape.

THURS

Objective Demonstrating that solids, liquids, and gases have weight.

Teacher Preparation Have available clothes hangers, empty milk cartons, paper cups, string, scissors, straws, pins, pictures or drawings of simply constructed scales, and at least one set of commercial balance scales.

Activities Working in pairs, students construct a weighing device of their own design. Once they have completed their scales, they weigh a variety of the solid objects that they worked with on Monday. They also weigh equal amounts of different liquids. Finally they weigh balloons and bags before and after they have inflated them. They determine a more precise measurement of weight by using commercial scales available in the classroom. They conclude that all forms of matter—solids, liquids, and gases—have weight.

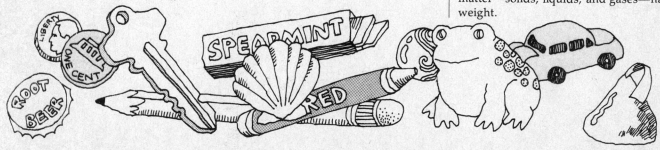

FRI

Objective Demonstrating that solids, liquids, and gases take up space.

Teacher Preparation Give each pair of students a cup, some marbles, a wooden block, paper towel, and access to water.

Activities Students count the number of marbles (or any other material) necessary to fill a glass or plastic cup. They empty the marbles and place a wooden block in the cup. They again count the marbles necessary to fill the cup and are asked to account for the difference—i.e., solids take up space. They empty the cup and refill it with water. They place the wooden block in the cup and observe that it does not go to the bottom of the cup but floats on the water that is now occupying the space in the cup. (Some objects will sink in liquids, displacing some of the liquid. The amount of liquid displaced is a measure of the amount or volume of both the liquid and solid involved.)

After drying the cup, they stuff a dry paper towel in the bottom and slowly immerse it upside down in a sink full of water (or aquarium or large bowl). After carefully removing the cup from the water, they remove the paper towel and observe that it is still dry. They conclude that air (gas) must be inside the cup and that it has kept the water out.

The week's activities are summarized by the concept that all forms of matter have weight and occupy space. Students have also seen that no two bits of matter can occupy the same space at the same time.

Notes

General Overview It is important for students to learn that the changes they see and experience in their natural environment are not entirely random and without cause. They need to understand some of the causes for change and the relatedness of certain changes. The activities of this week should help students gain insight into some of the more common phenomena in their environment: changing states of matter.

Learning Objectives Students will learn that matter can change from one form to another. They also learn that heating and cooling bring about changes in matter.

MON

Objective Observing and describing the changing of liquids to solids (setting or freezing).

Teacher Preparation Have some ice trays, water, a powdered mix such as Kool-Aid, and popsicle sticks available.

Activities Students fill some ice trays with water. They fill some other trays with a flavored mix and lay a stick in each cube. All of the trays are placed in the freezer compartment of a refrigerator. When frozen, students eat the popsicles and discuss what caused the change. They conclude that cooling caused a liquid to change to a solid.

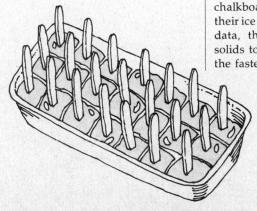

TUES

Objective Observing and describing the changing of solids to liquids (melting).

Teacher Preparation Give each student an ice cube on an aluminum pie pan or any other container.

Activities Half the students find a place in the classroom that will cause their ice cubes to melt the fastest. The other half find locations that will cause their ice cubes to remain frozen the longest. Make up a chalkboard chart with a place for students to write their name, location of the ice cube, and time it took to melt. Depending on the group, students choose locations near to or away from the windows and heat vents. They record on the chalkboard the amount of time it takes for their ice cubes to melt. From the summary data, they conclude that heat changes solids to liquids and the greater the heat the faster the change.

WED

Objective Observing and describing the changing of liquids to gases (boiling or evaporation).

Teacher Preparation Boil some water in a beaker on a hot plate, making sure to point out the level of water in the beaker before boiling.

Activities Students observe water as it begins to boil. They observe bubbles forming and rising to the surface and steam (water vapor) going into the air. As the boiling continues, the level of water in the breaker decreases visibly. They conclude that heating the liquid (water) changes it to a gas (water vapor).

THURS

Objective Observing and describing the changing of gases to liquids (condensation).

Teacher Preparation Again, boil some water in a beaker. Hold a pane of glass or glass lid over the steam.

Activities The students see that water vapor condenses on the glass held over the steam. If it is a cold day outside, they observe the condensation of water on the inside of the windows. In both cases, students conclude that the water vapor in the air changes to a liquid when it cools.

FRI

Objective Summarizing the relationship between changing states of matter and heating and cooling and applying knowledge of this relationship to practical events.

Teacher Preparation Draw a chart on the chalkboard with the following headings: *freezing, melting, boiling* or *evaporation,* and *condensation.* Also make copies of the puzzles shown in the activity below for each student.

Activities In small groups, students discuss some events involving changing states of matter for the teacher to record on the chalkboard. Their task is to describe events in which they have observed freezing, melting, boiling or evaporation, and condensation—for example, making ice cream, melting butter in cooking, visiting a hot springs or geyser, seeing moisture form on a bathroom mirror, etc. Still in small groups, the students try to complete the following puzzles and discuss their answers with the teacher.

1. Heating: __?__ → liquid → __?__
2. Cooling: gas → __?__ → __?__

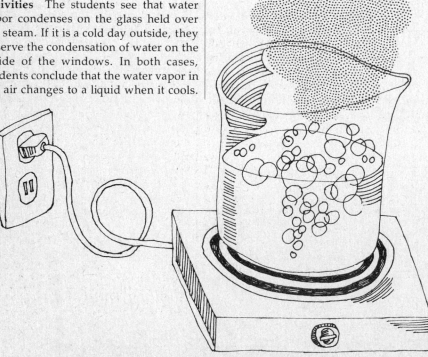

10

General Overview In addition to the changes in the states of matter studied in the previous week, matter can be described to change in two other ways: physically and chemically. All of these changes can be understood with greater clarity if one has knowledge of molecules and molecular motion.

Learning Objectives Students will learn that matter can undergo physical changes that alter the appearance of an object but do not produce a new material. They will also learn that matter can experience chemical changes that produce new materials. They will learn that matter is made up of smaller moving particles called molecules.

MON

Objective Observing and identifying various types of physical changes.

Teacher Preparation Provide each student with a small block of clay and later a small portion of hard chocolate. Have a pan and hot plate available.

Activities Students use the modeling clay to shape the form of an animal or some other object. They discuss how the clay has changed in shape but is still clay. Students then eat a small piece of a chocolate bar and melt the remaining portion on a hot plate. They observe that the chocolate changes from a solid to a liquid. When the chocolate has cooled somewhat, the students taste the melted chocolate to confirm that it is still chocolate. These two activities are identified as *physical* changes in matter, because no new material was produced. Students work with a partner and think of other physical changes. Their suggestions are listed on the chalkboard and discussed.

TUES

Objective Observing and identifying various types of chemical changes.

Teacher Preparation Provide students with a nail, a cup, water, and salt-free crackers. Have available scrap paper, matches, and a small container to burn the paper in.

Activities Students put a nail in a cup of water and observe the changes in the following days. Next they predict what will happen before the teacher burns a piece of paper, and then they observe and discuss the changes that do occur when the paper is burned. Next they chew on a salt-free cracker, delaying swallowing for as long as possible. They are asked to notice any change in taste (from starchy to slightly sweet). Each of these changes is identified as a chemical change, because new materials are produced: iron to rust (iron oxide), paper to ashes, and starch to sugar. Students work with partners and think of other chemical changes. Their suggestions are listed on the chalkboard and discussed.

WED

Objective Distinguishing between physical and chemical changes.

Teacher Preparation List a series of chemical and physical changes down the middle of a stencil. Label one side *physical change* and the other side *chemical change*. Label the activity the "Change Game." Distribute a copy to each student. Have available old magazines, scissors, paste, and tag board.

Activities Students identify each change —and any others they can add—as either chemical or physical by drawing an arrow to the appropriate side. As a group activity they make a "Change Collage," cutting pictures from papers and magazines and pasting them in groups according to the physical or chemical changes involved.

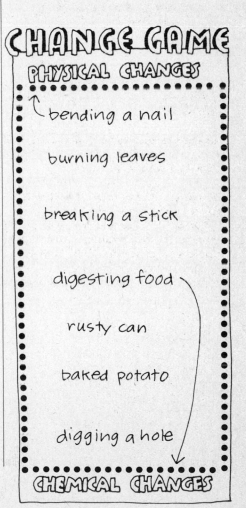

THURS

Objective Observing evidence for molecular structure and inferring that matter is made up of smaller particles called molecules.

Teacher Preparation Give each student a sugar cube, magnifier, drinking glass or cup, paper towel, and access to water. Introduce students to the word *molecule*, writing it on the board and discussing what it means.

Activities With the aid of a magnifying glass, students observe the characteristics of a sugar cube. They crumble the sugar cube into granules and observe these smaller components. They crush half of the granules to a fine powder to observe that sugar can be made into smaller and smaller particles. Next they fill a cup or glass of water to the brim and *slowly* pour the remaining sugar granules into the glass. They discover that even though the glass was "full," sugar could be added without spilling the water. The sugar molecules fill spaces between the water molecules. (This effect can be simulated by pouring sand into a cup full of marbles.)

Notes

FRI

Objective Observing evidence of molecular motion and inferring that changes in matter result from changes in the motion of molecules.

Teacher Preparation Open a bottle of perfume at one end of the room. Next drop dark food coloring or ink into a glass (or beaker) of hot water and a glass of cold water. Finally, heat an ice cube until it changes to water and then to steam.

Activities Students raise their hands when they smell the perfume fragrance, observing that those farthest from the perfume smell it last; they conclude that the perfume molecules move through air. When the teacher drops the food coloring into the glasses of cold and hot water, they observe the color mixing much faster in the hot water; they conclude that heat causes the molecules of food coloring to move faster. The students then discuss what is happening to water molecules as the teacher heats an ice cube to liquid and then allows it to boil and release steam; they should conclude that the changes in the states of matter (water) were caused by the water molecules moving faster.

6 MON

Objective Identifying and describing friction as the resistance or "holding back" effect between two objects rubbing together.

Teacher Preparation Have available some chalkboard erasers, rubber erasers, sponges, old books, rulers, coins or metal washers, and other rough and smooth objects.

Activities Students try to slide the various objects on different surfaces such as table tops, tile, wooden and carpeted floors, and concrete sidewalks. They observe that roughness affects how things slide. The teacher identifies the "holding back" of objects as *friction*—the same friction that students feel as heat when rubbing their hands together. Students observe that smooth objects slide easily on smooth surfaces because of less friction; rough objects are less easy to slide, especially on rough surfaces, because they produce more friction.

General Overview Friction affects virtually everything in the material world. Helping students understand friction and its relationship to their daily lives is an important goal. The lessons during week 6 are designed to introduce students to some of the more meaningful effects of friction.

Learning Objectives Students will describe friction and identify some ways in which friction is increased or decreased. They will also list some useful and harmful effects of friction.

TUES

Objective Identifying rollers, wheels, and ball bearings as ways of making objects move more easily (reducing friction).

Teacher Preparation Provide some round pencils or wood dowels (from a tinker-toy set), small wagon or other large toy with wheels, and heavy books or bricks.

Activities Students compare how easily an object such as a heavy book or brick moves across the floor under different conditions: first, when they put the object in a wagon and push the wagon; second, when they place the object on top of some round pencils or dowels layed out in parallel rows on the floor; and third, when they simply push the object on the floor. They observe that the movement of the book or brick is easier when pushed in the wagon or on the pencils than when flat on the floor. The wheels and rollers reduce the resistance (friction) on the moving object.

WED

Objective Using slippery materials (lubricants) to reduce friction.

Teacher Preparation Solicit student help in bringing some toys and skates with squeaking wheels to class. Locate some sticking hinges in the classroom and around the school. Have some lubricating oil available.

Activities Students push their squeaking toys to observe how well or poorly they roll. They put some oil on the wheels and bearings and push them again. They observe how much more smoothly the toys roll because resistance or roughness of friction has been reduced by the oil. They feel the slippery texture of the oil. Students then locate some squeaking hinges in the classroom and around the school and place oil on the hinges, observing that the noise is reduced and the hinges work more smoothly. They try to identify other slippery materials that could be used to reduce friction, such as graphite, petroleum jelly, and talcum powder.

THURS

Objective Identifying the unfair advantages in a "friction contest."

Teacher Preparation Provide a rope at least 6 meters (about 6 yards) long, two cardboard boxes filled with books or bricks, another cardboard box, a small wagon, and at least four jars with screw-on lids. Engage the students in a "friction contest," asking them to identify how increasing or reducing friction is used to make the contest unfair.

Activities In each contest the students identify how friction is used to give unfair advantage to one team. Contest 1 involves pushing a heavy box a given distance as quickly as possible. One student pushes the box across a polished wood or tile floor, and the other student has to push the box on a rough concrete walkway. Contest 2 involves one student pushing another student who is inside a cardboard box across the room faster than the opponent in the competing team. The competing team, however, has the student inside a small wagon instead of a box. The aim of Contest 3 is to unscrew the lids from two jars faster than the opponent. One student must put petroleum jelly on the palms of the hands, while the other student does not. Contest 4 is a tug of war with a rope, done on a smooth floor. One side wears shoes with rubber soles, and the other team wears only socks on their feet.

FRI

Objective Comparing useful and harmful effects of friction.

Teacher Preparation Ask the students why a bar of soap left on the bottom of a bathtub can be dangerous or why good brakes on a bicycle can be helpful. Relate these examples to friction. Ask them to look through some catalogs and around the room for examples of how reducing or increasing friction is helpful or harmful.

Activities Students identify and discuss ways in which reducing and increasing friction can be useful for fun, work, comfort, and transportation. They also discuss ways in which reducing and increasing friction can be harmful. The teacher lists their suggestions on the board.

Notes

General Overview Heat can be transferred from one material to another, but the rate of transfer and the reactions of materials to a transfer of heat are very different. Knowing this will allow students to utilize heat and materials in a more efficient and safe manner.

Learning Objectives Students will learn that heat can be transferred from one substance to another. They will learn that different materials react to heat in different ways. They also observe sources of waste heat.

MON

Objective Observing evidence that heat can be transferred from one substance to another.

Teacher Preparation Heat some water in a large pan. Put the end of a balloon over the mouth of a pop bottle and place the bottle in the pan of hot water. Turn off the heat source. Put another balloon over a different pop bottle and do not place it in the water.

Activities Students watch as the balloon slowly inflates over the bottle in the hot water, while the balloon over the second bottle remains deflated. They conclude that some of the heat from the hot plate is transferred to the pan, to the water, to the bottle, and on to the air in the bottle. The heated air "gets bigger" (expands) and inflates the balloon.

TUES

Objective Identifying materials that are conductors and nonconductors of heat.

Teacher Preparation Heat some water in a saucepan and place metal, wooden, glass, and plastic spoons in the water.

Activities Students take turns feeling the handles of the four different spoons and observe that the metal and glass spoons are much warmer than the wooden and plastic spoons. Metal and glass are identified as heat *conductors*, while wood and plastic are seen to be nonconductors (poor conductors) of heat.

SAVE THE ICE

Name	Time ice melted	Materials used

WED

Objective   Observing that light-colored materials reflect heat and dark-colored materials absorb heat.

Teacher Preparation   Set up two experiments: (1) Wrap the sides and bottom of a can with black paper, wrap another with white paper, and place thermometers inside both cans. (2) Put thermometers into two glasses of water, one of which is completely covered with a dark-colored cloth and the other with a very light-colored cloth. Place the cans and glasses for both experiments in full sunlight.

Activities   Students assist the teacher in setting up the experiments. They record the temperatures in the cans and the glasses of water when they first put them in sunlight and again three or four times throughout the day. They conclude that light-colored materials *reflect* heat better than dark-colored materials and that dark materials or objects *absorb* heat better than light-colored materials or objects. The teacher writes *reflect* and *absorb* on the board and discusses the importance of the findings to the students' daily lives—for example, selection of clothes, choosing paint for a house or car, etc.

THURS

Objective   Identifying the insulating effects of certain materials.

Teacher Preparation   Have available a large variety of materials: different colors of cloth patches, cardboard boxes, sand, sawdust, soil, aluminum foil, clear plastic cups, drinking glasses, paper cups. Show these materials to the students and then give each student an ice cube in a small pot-pie pan. Ask students to predict which material would keep the ice from melting the longest. Everyone must place their plates with the ice on the window sill.

Activities   Students test their predictions by selecting one material to prevent their ice cube from melting. They check their ice cube every ten minutes to see if it has completely melted. The melting times are recorded on the board. Students discuss their findings and identify those materials that best slowed the melting time. These materials are described by the teacher as better *insulators* than the other materials. They repeat the activity, this time trying to apply what they have learned.

FRI

Objective   Observing waste heat and predicting how it affects our surroundings.

Teacher Preparation   Take the students to different locations in and around the school where they can feel heat given off from machines, engines, appliances, and lights.

Activities   Students feel the heat being given off from a refrigerator, air conditioner, movie projector, electric mixer, and automobile engine. Ask them to discuss where the heat goes and what effect it has on our surroundings. They compare the amount of heat given off from an incandescent light bulb with that of a fluorescent light and describe which one is less wasteful of heat energy. Ask them to identify other sources of heat waste. Ask them to suggest ways of reducing the amount of wasted heat energy in their surroundings.

Notes

Learning about Light

General Overview Light is a form of energy that interacts with our environment in many ways. An understanding of some of these interactions contributes to a safer and more efficient use of light energy.

Learning Objectives Students will determine the necessity of light for human vision. They will also observe the interaction of light with different materials.

MON

Objective Observing that light is necessary for human vision.

Teacher Preparation Locate a room without windows that is completely dark with the lights off. Have the students sit around the room and describe the major objects and features in the room. Suggest that they quietly observe what they see when the lights are turned out briefly, then when a flashlight is turned on, and finally when the full lights are returned. (Students who have genuine fear of total darkness should not be asked to participate.)

Activities Students observe the room carefully with the room lights on. With the lights out, they can see nothing in the room. When the flashlight is turned on, they see only the people and objects near the light. The full room is visible again once the full lights are turned on. They conclude that light is necessary for sight.

TUES

Objective Identifying and classifying some of the sources of light.

Teacher Preparation With students participating, make a list on the board of some things that make light. After the idea is established, ask students to discuss in small groups other things that make light. Identify three major sources: *radiant* —sun, stars; *electrical*—light bulbs, lightning; and *chemical*—campfires, candles.

Activities Students identify some things that make light. They suggest such things as the sun, candles, light bulbs, neon signs, fireflies, lanterns, flares, lightning, campfires, fireworks, stars, etc. (If any students suggest the moon as a *source* of light, identify this as an example of *reflected* light. Reflected light is the focus of week 9. The moon's light is also the source of activity during week 18.) Students try to group their suggestions according to whether they involve radiant energy, electrical energy, or chemical energy.

WED

Objective Identifying some helpful and harmful effects of radiant energy.

Teacher Preparation Divide the class into two groups. Designate one group as the "Helpful" group and other as the "Harmful" group. Using the list of sources of light from the previous lesson, write the names one at a time on the board and ask each group to name one way in which the source could be helpful or harmful.

Activities The two student groups try to provide helpful and harmful effects of each light source. For the sun, for example, a helpful result is stimulation of plant growth; a harmful result is sunburn.

CAN THE LIGHT PASS THROUGH?

Yes	some	no
(transparent) plastic cup	(translucent) frosted glass	(opaque) aluminum foil

THURS

Objective   Observing that light passes through some materials and is stopped by other materials.

Teacher Preparation   Give students strips of aluminum foil, frosted glass, and clear plastic or glass. Ask them to hold these objects up to the room lights. Make stencil copies like the illustration here. Identify the various materials as being transparent, translucent, or opaque, but do not overemphasize these terms.

Activities   Students observe that: no light passes through the foil and they cannot see anything behind the foil; some light passes through the frosted glass and only fuzzy images can be seen; most of the light passes through the glass. They classify other objects in the room according to categories on their stencil worksheet. The teacher writes their suggestions on the board and they copy them onto their worksheet.

FRI

Objective   Making shadows with various materials.

Teacher Preparation   Make a three-paneled screen by cutting out the bottom, top, and one side of a cardboard box. Place this cardboard screen on an overhead projector in a dark room. Place a variety of familiar objects, one at a time, on the projector, using the screen to prevent the students from seeing the object. Use items such as a pencil, plastic and wooden rulers, clear plastic cup, watch, ball, pieces of glass plate, etc. Let the students guess the object. On a sunny day, take them outside to play shadow games.

Activities   Students look at the projected shadows and guess what the teacher has placed on the overhead projector. They observe that the thicker or darker (opaque) materials make the best shadows—even though they aren't always the easiest to guess. Outside they make their shadows appear to shake hands, dance, etc. without actually touching another person. They also make animal shapes with their hands. (They notice that light does not curve around materials or people.)

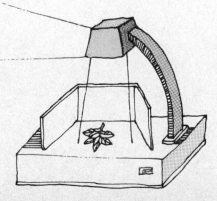

Bending Light

General Overview Light travels in straight lines even after its path has been deflected in some manner. This fact can be very useful if the different ways of bending light are understood.

Learning Objectives Students will observe that light travels in straight lines, but can be bent in several different ways. They will also learn that some materials reflect light more than others.

MON

Objective Testing the path of light.

Teacher Preparation Show the students a picture of a circus spotlight and ask them to observe the path of the light. Make the room as dark as possible and shine a bright flashlight around the room, asking the students to observe the beam of light. For the final activity, provide each student with a drinking straw to look through before and after bending.

Activities The students observe how a spotlight makes a straight beam in the sky. They observe the same direct path of light with the flashlight in the dark room. With the lights on again, they look through the straws. When they bend the straw, no light comes through and they see no objects. They conclude that light travels in straight lines. Ask them to think back to the shadow activities. If light did not travel in straight lines, there would be no shadows.

TUES

Objective Using mirrors to change the direction of light.

Teacher Preparation Solicit the students' help in bringing small mirrors to class. Give them ample time to observe how light is bent (*reflected*) by mirrors. Reinforce their observations that light continues to travel in straight lines even though it is reflected.

Activities Students observe themselves with their mirrors. They touch their faces to see how their movements are reflected by the mirror. Sitting in one place, they use the mirror to observe the objects in the room above, below, and behind them. They hold the mirror at an angle in the doorway to see down the hall. They share mirrors to bend light more than once, using the mirrors like a periscope.

WED

Objective Describing materials, textures, and colors that reflect light.

Teacher Preparation Make available a variety of materials that display characteristics such as smooth, rough, hard, soft, and light and dark colors. Darken the room and shine a flashlight on a sheet of smooth aluminum foil. Then shine the flashlight on a black piece of paper. Repeat this procedure for other materials.

Activities Students should observe the light reflecting off the aluminum foil, but very little off the black paper. After continuous observation, they conclude that smooth, shiny, and light-colored materials reflect light well, whereas rough, dull, dark-colored materials do not reflect light very well. They apply their knowledge to practical situations—for example, the use of black smut beneath the eyes of baseball players and black-colored windshield wipers on cars to reduce reflected light.

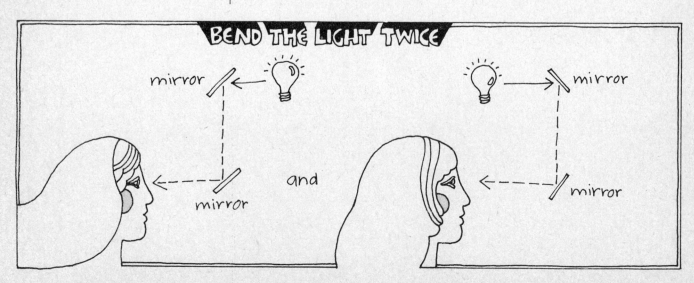

TOUCH THE MARBLE

THURS

FRI

Objective Bending light in water.

Teacher Preparation Give each student a clear plastic cup, marble, and pencil. Instruct them to place the marble in the bottom of the cup and touch it with the pencil, before and after filling the cup with water.

Activities Students observe that it is easier to touch the marble with the pencil when the cup is empty than when filled with water. The light reflected from the pencil travels differently in air and water and the pencil appears to bend.

Objective Using lenses to bend light.

Teacher Preparation Make available some magnifying glasses and other *convex* lenses for use in small groups. Have available a filmstrip projector, camera, microscopes, and other machines and objects that use lenses.

Activities Students use the magnifying glasses and convex lenses to observe how light is affected when it passes through the lenses. They can focus the light of the sun to produce heat. They can also magnify objects. They identify the convex lenses in a variety of useful machines and objects. They suggest other uses of convex lenses for glasses, telescopes, binoculars, gunsights, film projectors, etc.

CONVEX LENS

Observing Some Characteristics of Color

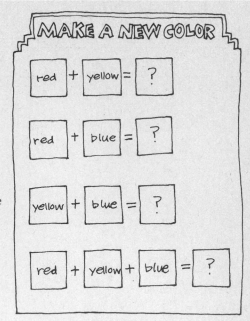

MAKE A NEW COLOR

red + yellow = ?

red + blue = ?

yellow + blue = ?

red + yellow + blue = ?

General Overview   Students live in a world of colors. Knowing that colors can be changed and used is very useful, because they affect moods, safety, and buying habits. An understanding of color can begin early in elementary school.

Learning Objectives   Students will learn that white light is composed of other colors, called its *spectrum*. They observe the effects of light passing through different colors. They also use the primary colors to make new colors. They discover that some colors are more visible than others and how helpful this is to them.

MON

Objective   Separating white light into different colors (its *spectrum*).

Teacher Preparation   Darken the room and project sunlight or slide-projector light through a triangular *prism* into a white wall or screen.

Activities   Students observe that a prism can separate white light into colors. They identify the colors projected on the screen. They infer that white light must be made up of colors they see projected (the *spectrum*). Students look for spectrums in water drops, windowpanes, magnifying glasses, etc.

TUES

Objective   Observing colors and their changes when light passes through them.

Teacher Preparation   Solicit student help in gathering some clear glass jars with lids. Divide the class into groups of three and give each group three jars of colored water: one red, one green, and one blue. Food coloring in water will suffice.

Activities   Students hold the jars of colored water in front of each other and observe the new color seen through the two jars. Ask them to overlap all combinations. Also ask them to reverse the positions of the jars to see if the colors change.

WED

Objective   Mixing primary colors of paints and observing new colors and black.

Teacher Preparation   Solicit student help in obtaining some jars—baby-food jars are very desirable for this activity. Use tempera to make separate jars of red, yellow, and blue colors, which are the primary colors of paint. Provide stirring sticks, paint brushes, and extra jars. Let the students test the colors produced when mixing the paint. Instruct them to save their paint for the next lesson.

Activities   Students mix the paints in different combinations to produce other colors. They paint the new color produced on the question jar on the worksheet (see example of worksheet). They discover that black is the product of mixing all colors.

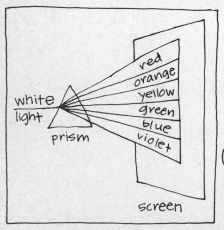

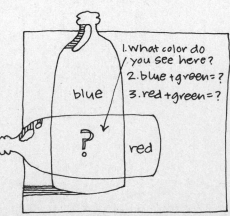

THURS

Objective Testing the visibility of different colors.

Teacher Preparation Cut some white paper or notebook paper into four equal parts. Provide the students with their paint from the previous lesson and some paint brushes.

Activities Students paint each piece of paper a different color. Half of the students then go to one side of the playground and the other half to the opposite side. On a signal from the teacher, they hold up their painted paper, one color (per student) at a time. No attempt is made to hold up all reds, then all blues, etc.; instead the colors are displayed in a random manner. The students observe which colors are more easily seen (more *visible*) from a distance.

General Overview Electricity is one of the major forms of energy used in our society. Knowledge of its source, its use, and its safe application are useful to the everyday lives of most people, which is the focus of the eleventh week.

Learning Objectives Students will name and identify some of the uses and sources of electrical energy. They will describe some of the safety rules that apply to electricity.

FRI

Objective Describing different uses of color visibility to make our lives safer.

Teacher Preparation Show the students the red warning label on some package or bottle and a brightly colored life jacket. Point out the color of a nearby "Exit" sign. Discuss the importance of using more visible colors at night when walking or riding a bicycle.

Activities Students describe other uses of color visibility in making their lives safer: at school, at home, riding to school, crossing a street, boating, hunting, etc.

MON

Objective Identifying some of the uses of electrical energy.

Teacher Preparation Ask the students to name some ways in which electrical energy works for them. Try to categorize their initial responses into groups such as: heating and cooling, motion and power, lighting and communication. Write these categories on separate pieces of posterboard. Provide the students with old newspapers and magazines, scissors, and paste.

Activities Students locate pictures of other illustrations of the uses of electrical energy in the newspapers and magazines. They cut and paste their pictures on the appropriate posterboard.

TUES

Objective Identifying sources of current electricity.

Teacher Preparation Show the students some examples of the more common sources of chemical electrical energy: dry cells such as flashlight batteries and storage batteries such as used in automobiles. Also show them a picture of a power plant, which generates mechanical electrical energy. Automobiles also have generators that produce electricity.

Activities Students look again at their posterboard collage of electrical machines, appliances, tools, toys, communication equipment, transportation devices, etc. They discuss the source of power for each object. They identify some objects such as radios that are powered both by dry-cell batteries and electricity from power generators.

Notes

WED

Objective Identifying sources of static electricity.

Teacher Preparation Ask the students if they have ever heard the term "electrical storm" used when lightning was flashing. Identify this as a form of *static electricity*. Provide each pair of students with a balloon and piece of wool or fur (or sweater).

Activities Students rub the inflated balloon with a piece of wool. They observe what happens when they hold the balloon next to their hair. They observe the balloon stick to the wall. The teacher identifies this as *static electricity* also. Students recall other times they have observed evidence of static electricity, such as brushing hair or walking across a carpet and touching a door knob.

THURS

Objective Describing the effects of electric current on different types of copper wire.

Teacher Preparation Provide each pair of students with a battery and three different types of copper wire: a thick, bare copper wire (#20), a thin, bare copper wire (#36), and a plastic-covered copper wire (#20). Following the activity, discuss the danger of bare wires in a house and wires or cords that are too small for the amount of energy.

Activities Students connect, in turn, the three different pieces of wire to the positive and negative ends of a battery (complete a circuit) and feel the heat from the wires. (The thin, bare wire will become warm rather quickly.) They hold two batteries together to increase the amount of electricity flowing through the wires and again feel the heat produced. (Hold the wires to the batteries for only a few moments; otherwise the battery is used up quickly.)

—NOTICE—
Strip the coating back from the ends of the plastic-covered wires where it touches the battery.

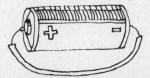

FRI

Objective Identifying and describing safety rules for electricity.

Teacher Preparation Show students the warning labels on some electrical tool or in a magazine picture. Discuss some of their experiences with electricity when caution was not observed. Propose the development of a list of safety rules for electricity, such as:

1. Do not put your finger into an electrical socket.
2. Replace worn or frayed cords on appliances.
3. Don't stand under trees during a lightning storm.
4. _____.
5. _____.

Activities Students use their own ideas as well as those they discover from books, magazines, and talking with other people to make a list of safety rules to display in class.

Notes

Making Circuits, Conductors, and Electromagnets

General Overview A minimal understanding of electricity is almost mandatory in order to function as a scientifically literate person in today's world. Though scientists still have questions concerning the exact nature of electricity, it can nevertheless be predicted to function in specific ways. Some of the ways in which electricity behaves are the focus of the activities of the eleventh week.

Learning Objectives Students will learn how to make electrical circuits. They will discover materials that are conductors and nonconductors of electricity and apply this knowledge in making a switch. They will also learn how to make a magnet using electricity.

MON

Objective Producing a single electric circuit.

Teacher Preparation Give each student a fresh flashlight battery (C and D cell batteries work well), a flashlight bulb, and about 15 cm (about 6 inches) of thin, bare copper wire (#36 works well). Instruct the students to use the materials provided to light the bulb. Encourage them to share their methods. Identify those that work as complete *circuits*.

Activities Students attempt a variety of arrangements to light the bulb. After a period of time, several different ways will be found to work. They find that the top (+) and bottom (−) of the battery must be touching either the wire or bulb and that the side *and* bottom of the metal part of the bulb must be touching either the battery or wire.

TUES

Objective Enlarging the circuit.

Teacher Preparation Let the students work individually or in small groups in trying to complete a variety of tasks requiring that they increase the number of parts in their electrical circuits. Have a sufficient supply of batteries, thin copper wire, and bulbs available.

Activities Students attempt to complete the task provided by the teacher. They also make up their own variations and draw them on the board. They observe the effects of the different approaches on the amount of light produced by the bulb.

WED

Objective Identifying conductors and nonconductors of electricity.

Teacher Preparation Have available some wood splints or popsicle sticks, metal and plastic spoons, aluminum foil, pencil erasers, string, and glass (microscope slides work well). Also have some batteries, copper wire, and bulbs available. Tell students that their task is to use the materials in trying to complete a circuit. Provide a stencil worksheet to write or draw their results, such as the example shown.

Activities Students select a number of objects to use one at a time in trying to complete an electrical circuit. They use the materials provided by the teacher and any others in the room. They summarize their results on a worksheet and discuss the meaning of *conductors* and *nonconductors*.

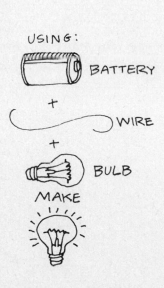

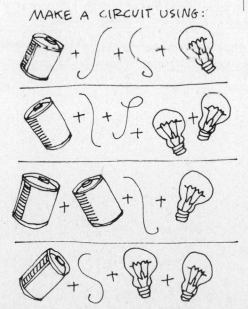

COMPLETING A CIRCUIT	
CONDUCTORS	NONCONDUCTORS
metal spoon friendship ring (etc.)	popsicle stick shoelace (etc.)

THURS

Objective Making an electrical switch.

Teacher Preparation Solicit student help in cutting strips from the bottom of aluminum pie pans about 5 cm (2 inches) wide. Cut the strips in half so that each pair of students would be given two strips. Have available some cardboard (old boxes), batteries, bulbs, and 15 cm (about 6 inches) of copper wire. Provide a diagram for making a switch.

Activities Student groups construct a switch. When they connect the two strips of aluminum (a conductor), the light comes on; when the strips are not touching, the light is off.

FRI

Objective Using electricity to make a magnet.

Teacher Preparation Provide each pair of students with a battery, large nail, and 30 cm (12 inches) of plastic-coated copper wire (#20 works well). Scrape the ends of the wire where it attaches to the battery. Provide a variety of materials such as paper clips, staples, iron filings, string, eraser, and pencil for them to test.

Activities Students construct an *electromagnet* like the one illustrated and observe and test their magnet on a variety of materials. (Hold the wire to the battery only briefly.)

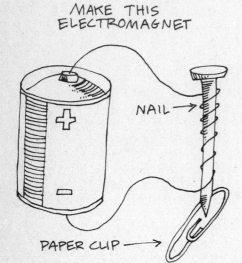

MAKE THIS
ELECTROMAGNET

NAIL →

PAPER CLIP →

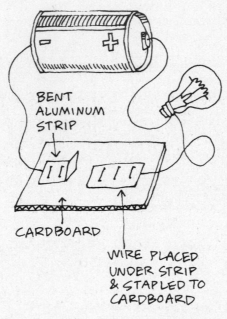

BENT
ALUMINUM
STRIP

CARDBOARD

WIRE PLACED
UNDER STRIP
& STAPLED TO
CARDBOARD

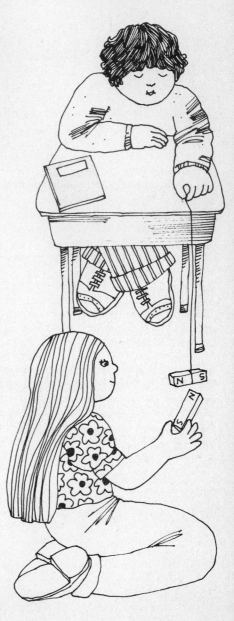

Notes

24

Describing Magnets and Magnetism

General Overview Magnets are materials that will attract materials made of iron, steel, cobalt, and nickel. The American nickel coin is mostly copper and is not attracted by magnets. The force (push or pull) of a magnet is strongest at its ends, called *poles*. Manufactured magnets have two poles: a north-seeking pole and a south-seeking pole. When two magnets are brought together, the two like poles repel each other and the unlike poles attract each other. Magnetic forces can pass through nonmagnetic materials.

Learning Objectives Students learn what kinds of materials can be attracted by magnets even through nonmagnetic materials. They also observe evidence for laws of magnetic attraction and magnetic fields.

MON

Objective Determining what materials are attracted by magnets.

Teacher Preparation Make available a variety of materials such as paper clips, rubber bands, wood, chalk, paper, glass, aluminum foil, rocks, and nails. Give each student or pair of students a magnet. You may want to let them use two different types of magnets: the bar and horseshoe (U-shaped) magnets.

Activities Students use the magnet to test what materials are or are not attracted by their magnets. They classify their finding into two groups. They test various fixed objects around the room such as the door handles, windows, sink, chalkboard, etc.

MAGNETS	
will attract	will not attract
paper clips	wood
nails	chalk
(etc.)	(etc.)

TUES

Objective Discovering that magnets can attract through nonmagnetic materials.

Teacher Preparation Have available some of the materials found to be magnetic in the previous lesson. Also provide a paper cup, plastic cup, drinking glass, aluminum pie pans, and bar magnet.

Activities Students place some materials that can be attracted by magnets in the paper cup, plastic cup, drinking glass, and aluminum pie pan. They bring a magnet close to the containers and observe any indication of attraction between the magnet and objects inside the containers.

WED

Objective Identifying the north-seeking and south-seeking poles of a magnet.

Teacher Preparation Give small groups of students a bar magnet and some string. Instruct them to tie the string around the magnet and adjust the position of the string to balance the magnet in a free-swinging position. Later give the students a small compass.

Activities Students hang their balanced magnets off the edge of a table or desk away from any metal. They observe that all the magnets in the room have the end marked "N" pointing in the same general direction. When the teacher gives them a compass, they observe that the "N" end of the magnets is pointing in the same direction as the compass needle: north. The ends of a magnet are identified as the north-seeking (N) pole and the south-seeking (S) pole.

THURS

Objective Discovering the law of magnetic attraction.

Teacher Preparation Give small groups of students two bar magnets and some string. Instruct them to tie the string around one of the magnets and adjust the position of the string to balance the magnet as they did in the previous lesson.

Activities Students hold the north-seeking pole of another magnet near the north-seeking pole of the suspended magnet and observe. They hold the south-seeking poles of the two magnets close together and observe. Then they hold the north-seeking end of the magnet in the hands near the south-seeking end of the hanging magnet. They observe that like poles *repel* each other and unlike poles *attract* each other.

FRI

Objective Observing a magnetic field.

Teacher Preparation Provide small groups of students with a bar magnet, a piece of white posterboard, and iron filings or small bits of fine steel wool.

Activities Students place a sheet of posterboard over a bar magnet. They sprinkle some iron filings over the posterboard and tap the posterboard several times. They observe the filings forming a definite pattern with a concentration at the poles.

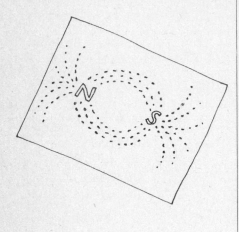

Notes

Learning about Sound

General Overview Sound is a special form of wave energy, and it affects students in many ways every day. It is the major form of communication. Even totally deaf students sense vibrations from some sounds. How sound is produced and used is the focus of this week's lesson.

Learning Objectives Students will learn that sound is produced by vibrating objects—matter moving back and forth rapidly. They also learn that sound can travel, and thus is a vital form of communication for humans and other animals. They also identify some of the important safety sounds in their environment.

MON

Objective Producing sounds from a variety of objects.

Teacher Preparation Solicit student help in collecting a large variety of objects such as aluminum pie plates, bottles, bottle caps, string, wire, rubber bands, wood of various sizes and shapes, tin cans, ice-cream buckets, old tire tube, scissors, nails, hammer, and saw. Virtually anything would be appropriate. Instruct them to use the materials to make a musical instrument.

Activities Students pick the objects they want to use and make one or more musical instruments. They use their instrument in playing along with a song. They discuss how each instrument produces sound.

TUES

Objective Discovering that sounds are produced by vibrations.

Teacher Preparation Strike a tuning fork and let the students take turns feeling the vibrating fork. Strike the fork again and place it in a bowl of water. Ask the students to hold their throats while they hum.

Activities Students hear the sound of the tuning fork and feel the vibrations. They also see the waves produced by the vibrating fork when it is placed in water. They hear the sound and feel the vibrations in their throats when they hum.

WED

Objective   Communicating by sound.

Teacher Preparation   Solicit the students' help in procuring enough empty tin cans (soup-can size) for everyone in the room. Make sure that all cans are clean and that they have no sharp edges where the top has been removed. Drive a nail through the bottom end of each can. Provide each pair of students with two cans, two buttons (the type without shanks), and six or more meters of string or wire.

Activities   Students pull the ends of the string through the holes in the cans and tie the ends to the buttons. They hold the cans far enough apart so that the string is pulled tight and the buttons are held flat to the inside bottom of the cans. They take turns talking into the cans and hearing each other's voices. They discuss how sound can travel and identify some daily experiences they have with sound traveling long distances—telephones, radios, television, etc.

THURS

Objective   Identifying the communication sounds of animals.

Teacher Preparation   Give each student a piece of paper with the name of a well-known animal. (Give each student a different animal name.) Instruct them not to tell anyone the animal name on their paper, but when called upon, to make a noise that sounds like their animal. Following this activity, take students to a wooded area and listen to animal sounds. (As a follow-up to this activity, play records of different animal sounds, take a field trip to an animal park or farm, etc.)

Activities   Each student tries to imitate the sounds of the animal named on their piece of paper. The other students try to identify the animal. They discuss some of the ways that animals make their sounds and why it is important that they make these sounds, such as defense and locating a mate. They hear the sounds of birds, insects, and perhaps frogs and squirrels on their field trip.

FRI

Objective   Identifying safety-oriented sounds.

Teacher Preparation   Collect a variety of pictures of objects associated with specific sounds as safety devices: police car, ambulance, fire truck, train, railroad-crossing signal, car horn, safety-belt and car-door buzzers, school fire warning, etc. Show the pictures to the students and ask them to recall the sounds suggested by the pictures. If possible, arrange a field trip to the fire and police stations for a siren demonstration.

Activities   Students imitate some of the "safety" sounds that they relate to the objects in the pictures. They discuss how the sounds are helpful to themselves and others. They recall other safety sounds: bike horn, civil-defense radio alert, time buzzers on stoves, etc.

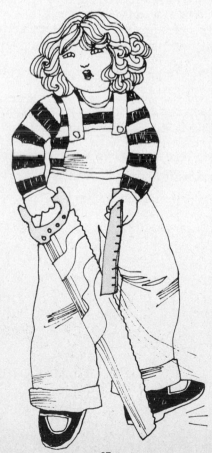

Notes

27

General Overview A practical knowledge of the effects of gravity is basic to the understanding of things as ordinary as indoor plumbing to something as extraordinary as space travel. Each lesson in the fifteenth week emphasizes different aspects of gravity.

Learning Objectives Students will observe the effects of gravity on objects of different sizes and weights. They will also locate the center of gravity in objects.

MON

Objective Observing the effects of the earth's gravity.

Teacher Preparation Have rubber balls, Ping-Pong balls, and tennis balls available.

Activities Students take turns releasing the different balls and observing that they always fall toward the earth. They watch the teacher pitch the ball up and observe the ball again return to the ground. They identify or are told by the teacher that the earth's *gravity* is the force that pulls the objects down.

They select an object from their desk or pocket and release it, observing that their object and objects dropped by other students are pulled down by gravity. They conclude that gravity affects all objects on earth.

WED

Objective Locating the center of gravity in a variety of objects.

Teacher Preparation Provide each student with a ruler or meter stick or yard stick and a small lump of clay. Ask the students to balance their stick horizontally on the end of their fingers.

Activities Students balance their measuring sticks. They observe that the middle of the stick is the center of gravity. They put some clay on one end of their stick and balance the stick again. This time they observe that the center of gravity has shifted toward the end with the clay on it.

THURS

Objective Observing and inferring that weight is a measure of gravity.

Teacher Preparation Have available a set of bathroom scales and simple classroom balances.

Activities Students hold a book with their arm held straight out in front or beside them for as long as they can. The teacher identifies gravity as the constant force that makes the book feel heavier (causes the book to have weight). Students weigh themselves and a variety of classroom objects of different sizes and weights. They observe that bigger objects do not always weigh more than smaller objects, but *all* objects have weight, which is a measure of the earth's gravitational pull on the objects.

TUES

Objective Observing that gravity is the same around the earth.

Teacher Preparation Have available a world globe, filmstrips, and books with pictures of people and buildings in different parts of the world.

Activities Students observe pictures of people living in different parts of the world. With the teachers help, they locate on the globe some of the places referred to in the pictures. When asked if gravity is the same on the top, sides, and bottom of the earth, they conclude that it is since people, buildings, and trees all remain standing and do not "fall off the earth."

FRI

Objective Observing and inferring that the weight of an object changes depending upon its distance from the center of the earth.

Teacher Preparation Have available a film and/or pictures showing the weightless conditions of astronauts in space. Explain how very high above the earth spaceships are.

Activities Students observe the pictures of astronauts and objects floating in orbiting spacecraft. They conclude that the pull of gravity on the astronauts is less. Though some students suggest that there is no air to "hold them up," the teacher points out that there is air inside the spacecraft.

Notes

General Overview Though young students have limited knowledge and concepts of the solar system, they have heard and seen things about it on radio, television, and in newspapers. Students are usually interested to learn more about the solar system, which is the focus of activities in the sixteenth week.

Learning Objectives Students will describe the position, motion, and other information about the planets and other objects in the solar system. They will also identify the sun as the source of light in the solar system.

MON

Objective Describing the position and motion of planets in the solar system.

Teacher Preparation Draw a series of nine concentric circles spaced a few feet apart on the floor around the center area that is marked as the sun. Label nine cards with the names of the planets. Select students to stand on each of the circles in a straight line from the sun, in the order: Mercury, Venus, Earth, Mars, Jupiter, Saturn, Uranus, Neptune, and Pluto. Give each student the appropriate planet label. Ask them to walk at a steady pace on their circle in a counterclockwise direction around the sun.

Activities Students observe that the closer planets, Mercury and Venus, go around the sun quicker than the earth, and all the other planets take longer than the earth to go around the sun. Students take turns pretending to be planets. The teacher identifies the motion around the sun as the *revolution* of the planets.

TUES

Objective Observing the source of light and heat from planets in the solar system.

Teacher Preparation Divide students into groups of nine. Provide modeling clay for every student. Ask each student in the group to form their clay into a different planet, so that all planets are represented in each group. (Don't be concerned with the size of the planets here.) Place a floor lamp without the shade in the center of the room and draw nine circles around this "sun" representing the revolution of each of the planets. The sun is identified as a star and the main source of light and heat for all planets in the solar system.

Activities Students take turns standing on the circle that corresponds to the planet they have modeled. They hold their planet in front of them and revolve around the sun. They observe that the planet always gets some light from the sun no matter where it is in its orbit. By observing other students' planets, they notice that the planets closest to the sun get more light and are brighter, while the planets farther away from the sun are dimmer and receive less light and heat.

WED

Objective Describing features of the solar system.

Teacher Preparation Have available an abundance of books, films, and filmstrips about the solar system. Label a series of envelopes with the names of members of the solar system: the sun, each of the planets, comets, and meteorites, plus an envelope for general information about the solar system. Assist the students in making a series of fact cards about the solar system.

Activities Using the books, films, and filmstrips about the solar system as their sources, students identify information that can be placed on fact cards, such as: Saturn has rings, Mercury is the nearest planet to the sun, shooting stars are really meteorites, and all objects in the solar system orbit around the sun. Each fact card is placed in the appropriate envelope. The teacher reviews the fact cards with the students. They also play "Solar System Jeopardy" with the cards, using the information on the cards to make up questions and let other students guess the answers.

SOLAR SYSTEM DIAMETERS

Scale: 1 cm = 10,000 miles

Sun =	86 cm (34 in.*)
Mercury =	3 mm (⅛ in.)
Venus =	8 mm (¼ in.)
Earth =	8 mm (¼ in.)
Mars =	4 mm (⅛ in.)
Jupiter =	8.7 cm (3½ in.)
Saturn =	7 cm (2¾ in.)
Uranus =	3 cm (1¼ in.)
Neptune =	3 cm (1¼ in.)
Pluto =	4 cm (⅛ in.)

*inches approximated

DISTANCES FROM THE SUN

Scale: 1 meter = 10 million miles

Mercury =	3½ meters
Venus =	6½ meters
Earth =	9 meters
Mars =	14 meters
Jupiter =	48 meters
Saturn =	88 meters
Uranus =	178 meters
Neptune =	280 meters
Pluto =	367 meters

THURS

Objective Observing the relative sizes of the planets in the solar system.

Teacher Preparation Provide the measurements for a scale model of the sizes of the planets in the solar system. Depending on the skills of the students, either provide them with the rulers and compasses to make the drawings or give them line drawings to trace. Have scissors available.

Activities Students draw or trace the sizes of the different planets according to the scale provided by the teacher. They cut out the circles representing the planets and place them next to each other and on top of each other in various combinations in order to compare the relative sizes of the planets.

FRI

Objective Modeling distances in the solar system.

Teacher Preparation Provide the measurements for a scale model of average distances from the sun in the solar system. Have meter sticks available.

Activities With the teacher's help and with a scale of distances provided by the teacher, students measure and mark the straight-line distances between the sun and the planets on the school ground. Students representing the planets stand in the marked locations, holding a sign with the name of the planet that they represent. Other students walk the entire scaled distance between the sun and planets.

Notes

Describing Some Effects of the Earth's Motions

General Overview Understanding how the earth moves in space is not easily observed in nature. The earth's motion can be explained, however, by using models and then relating the models to what can be observed in nature. Each lesson during the seventeenth week emphasizes the earth's two major movements in space, rotation and revolution, and how they affect the apparent motion and position of the sun.

Learning Objectives Students will identify and describe the motion of the earth that causes day and night and seasons. Students will also relate the earth's revolution to changes in the sun's rising and setting points and to shadows produced by the sun.

MON

Objective Explaining day and night by simulating the earth's rotation.

Teacher Preparation Provide each student with some clay and ask them to form it into a ball. Place a lamp (without the shade) in the middle of the room. Turn out the overhead lights and close the window shades. Tell the students that the lamp represents the sun and their clay balls represent the earth. Ask them what causes day and night.

Activities Students observe that the light shines on half of their clay ball no matter how far it is from the light and no matter how they turn it. They push a pencil through the middle of the clay to simulate the earth's *axis*, describing the top of the pencil as north and the sharp end as south. They turn it counterclockwise (as demonstrated by the teacher) to simulate the earth's rotation on its axis. They identify the light side as day and the dark side as night. Sunrise and sunset lines are identified by marking a spot on the clay and turning the ball one complete rotation so that the spot passes from "night" to "day" and into "night" again. The teacher demonstrates day and night on a world globe.

TUES

Objective Defining a year as one complete revolution of the earth around the sun.

Teacher Preparation With chalk, draw an elliptical path around a floor lamp placed in the middle of the room. Ask the students to take turns walking around the path in a counterclockwise direction.

Activities Students walk along the elliptical path until they return to their starting point. The teacher identifies their motion as one *revolution*, and points out that one revolution of the earth around the sun takes one year. Students are asked to walk the path again, but this time they should *rotate* as they *revolve* around the "sun." Ask them to recall the number of days in a year, 365, which is the number of times the earth rotates on its axis as it makes one revolution about the sun.

WED

Objective Explaining what causes the earth's seasons.

Teacher Preparation Use a world globe and a floor lamp in a darkened room to demonstrate the earth's 23½° tilt in relation to the sun. Keep the direction of the tilt constant and move the globe around the "sun" to simulate the positions of the earth at the beginning of summer, fall, winter, and spring.

Activities Students observe the areas on earth that receive the most direct solar rays year round (the equatorial zone), the areas that experience long periods of darkness and daylight in different seasons (polar regions), and the areas, such as those in which they live, that experience seasonal changes (temperate zones).

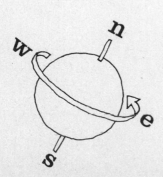

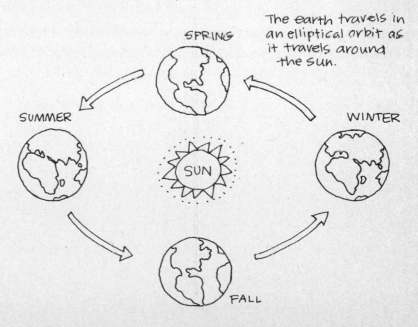

The earth travels in an elliptical orbit as it travels around the sun.

THURS

Objective Observing seasonal changes in the rising and setting positions of the sun.

Teacher Preparation Caution the students against looking directly at the sun. Provide students with a compass. Define *horizon* as the point where the sky and earth appear to meet—the limits of what they can see outside.

Activities At the beginning and end of the school day, students locate the sun's position in relation to objects near the horizon at a particular place on the school ground. With their compasses, they locate the general direction of the sun in the morning and afternoon. They return to the same place at the same time every week or so throughout the year to observe and record changes in the sun's location near the horizon. The teacher encourages them to observe the horizon position of the sun in the early morning and near sunset at their homes. They relate changes in the rising and setting positions of the sun to the earth's revolution and changes in the seasons.

Notes

FRI

Objective Relating seasonal changes in shadows to the sun's position in the sky.

Teacher Preparation Provide each student with a soda straw, lump of clay, and piece of paper.

Activities At the beginning of the school day, students put a lump of clay in the middle of a piece of paper on a sidewalk exposed to sunshine all day. They stick a soda straw (or any straight object) straight up in the clay. They draw a line from the shadow made by the straw and write the time of day on the line. They repeat this activity every hour throughout the day. They notice that the most direct rays of the sun occur at noon when the shadow is the shortest. They repeat the activity at least once a month throughout the year, comparing the lengths of the shadow and observing that the sun's position in the sky changes throughout the day and year. They relate changes in the length and position of shadows to the earth's revolution and changes in the seasons.

Describing the Moon and Its Motions

General Overview   The largest natural object in space that is nearest to the earth is our moon. It affects the earth more than any other celestial body except the sun. It is usually the most prominent object in our night sky and is the object of intense space exploration. Some of what is known about the moon and its motions can be understood in the elementary years.

Learning Objectives   Students will identify and describe some of the moon's features and motions. They will demonstrate the causes of the lunar phases and eclipses of the sun and moon.

MON

Objective   Identifying some of the features of our moon.

Teacher Preparation   Have available a variety of books, pamphlets, and articles containing information about the pictures of the moon. (NASA, Washington, D.C. 20546, is an excellent source for moon information and pictures.) Begin a discussion on what the students know or have heard about the moon from radio, television, school, or any other source. Make a list of moon facts on the board.

Activities   Students describe some of the things they know or have heard about the moon. Using the resources available in the room, they discover other facts about the moon: that it is one-fourth the size of the earth, has no air and water, has mountains, craters, old volcanos, and large flat areas called "seas." The teacher adds new information to the list of moon facts. Students begin a month-long moon watch, observing the moon's phases and trying to locate some of its more obvious surface features.

TUES

Objective   Making a scale model of the earth and moon.

Teacher Preparation   Provide each student with some modeling clay and rulers. Provide a convenient scale to use in making a model of the earth and moon, 2.5 cm or 1 inch = 3,200 kilometers or about 2,000 miles.

Activities   Students make a scale model of the earth and moon with the clay. The earth is 10 cm (4 in.) in diameter, and the moon is 2.5 cm (1 in.). They place the clay balls 305 cm (10 ft) apart.

WED

Objective   Describing the motion of the moon around the earth.

Teacher Preparation   Provide students with modeling clay. Ask them to mold the clay into a ball about the size of a tennis ball.

Activities   Working in pairs, one student remains seated and represents the earth. The second student marks a spot on the clay that represents the moon and walks counterclockwise around the first student, always holding the marked spot facing the seated student. They reverse roles and repeat the activity. They observe that the moon rotates once as it revolves once around the earth, thus only one side of the moon can ever be seen.

THE MOON PHASES

as seen from earth:

as seen from space above the earth:

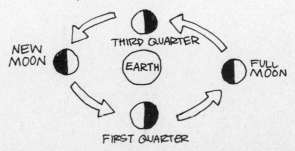

THURS

FRI

Objective Simulating the phases of the moon.

Teacher Preparation Provide student pairs with their clay moons (or styrofoam balls). Place a lamp, with the shade removed, in the middle of the room to represent sunlight. Darken all other lights in the room.

Activities Student pairs repeat the procedure in the previous activity, but this time they observe carefully how much of the lighted part of the moon they can see. Identify moonlight as reflected sunlight. Caution the students against blocking the light from the "sun" with their bodies while rotating and revolving the "moon." They are able to observe a simulation of the moons phases, easily identifying the full, new, and quarter moons. The crescent moons are seen between the new and quarter moons.

Objective Describing eclipses and how they occur.

Teacher Preparation Again, place a lamp, without the shade, in the middle of the room and provide students with their clay moons. Place one or more globes a meter (about a yard) or two from the lamp. Write the word *eclipse* on the board. Ask students if they know what the word means. Darken the room except for the floor lamp.

Activities Students put their clay "moons" directly between the light and the globe, the position of the new moon. They can observe a shadow being cast into the earth by the moon. People at that spot would not be able to see the sun because of a *solar eclipse*. When students position the moon at full moon, with the globe directly between the sun and moon, the moon will be in the shadow of the earth (its light cut off by the earth) and will not be visible, illustrating a *lunar eclipse*. The teacher explains that eclipses do not happen every new and full moon, because the moon's orbit around the earth is tilted—it doesn't always move directly in front of and behind the earth, relative to the sun.

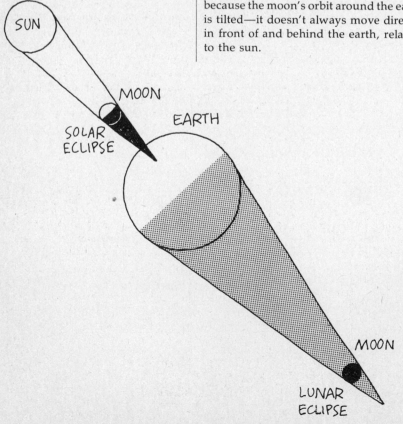

34

General Overview The interacting forces and conditions that affect the earth's weather are exceedingly complex. There are some basic processes, however—such as heating and cooling of soil, water, and air—that are relatively easy to comprehend and lead to a general understanding of weather. The processes are introduced in the lessons in the nineteenth week.

Learning Objectives Students will identify the effects of the earth's tilt and motions on temperature changes in soil, water, and air. They also observe in more detail how heating and cooling affects air and creates wind.

MON

Objective Describing the effect of day and night on the heating and cooling of soil, water, and air.

Teacher Preparation Have available at least three cans or jars of equal size and shape, three thermometers, soil, and water.

Activities Students fill one jar with soil, one with water, and leave one filled with air. They place thermometers inside all jars and place them outside in clear weather for twenty-four hours, recording temperature readings every hour throughout the school day and beginning early again on the second school day. They discover that the temperatures of the soil, water, and air increase and decrease as a day and night pass, but that: (1) they do not rise and fall at the same rate, and (2) they do not reach the same temperatures.

TUES

Objective Relating the earth's tilt to differences in heat absorption.

Teacher Preparation Have available two thermometers and at least two shoe boxes of equal size—more if you want to actively involve the whole class. Using a globe, review with the class the information that the earth is tilted relative to the sun, that it turns on its axis (rotates) every twenty-four hours, and that it goes around the sun (revolves) once every year.

Activities Students fill the boxes with soil and insert thermometers to the same depth in the boxes. They let one box lie flat on the ground and prop up the other box so that its surface is facing the sun's rays more directly. They make temperature readings at least every half hour and record their results. They discover that the soil tilted toward the sun and receiving more direct sunlight heats faster than the soil lying flat and receiving the sun's rays at an angle. The teacher and students discuss the relationship between their findings and the earth's tilt.

WED

Objective Observing that hot air expands and cold air contracts.

Teacher Preparation Supply three balloons for each group of three students. Have access to a refrigerator. Explain the terms *contract* and *expand*.

Activities Students in the same group blow up all three balloons to as nearly the same size as possible and tie the ends. They wrap a piece of string around the middle of each balloon and cut the string at the point where it completes a circle (a measure of circumference). They place one balloon in a refrigerator, one in full sunshine, and one in the shadows of the classroom. After an hour has elapsed, they measure the balloons again with new string. They tape their strings on a chart and discuss their results. (All the balloons are likely to be smaller due to air leaking out; the balloon in the room acts as control for how much air leaks out.)

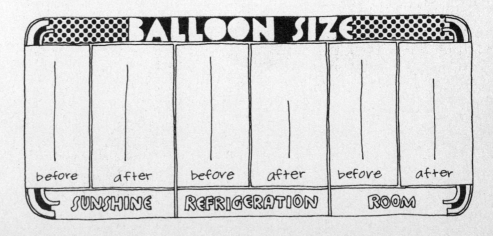

BALLOON SIZE

| before | after | before | after | before | after |
| SUNSHINE | | REFRIGERATION | | ROOM | |

THURS

FRI

Objective   Observing evidence of hot air rising and cold air sinking.

Teacher Preparation   Have available two down feathers, a flexible study lamp with a bright bulb, and access to a refrigerator. Identify *wind* as movement of air due to unequal heating.

Activities   Students hold feathers above and below the door of a freezer compartment. When the door opens the feather below the door flutters as the cold air rushes down over it. The feather held above the door shows no movement. Students also feel the cold air flowing down as the freezer door is opened.

Students hold feathers about 10 cm above and below a study lamp after the light bulb has been turned on for a few minutes. They observe that the feather held above the bulb flutters. They also feel more heat above the bulb than below it.

Students conclude from the two activities that cold air is heavier and falls and hot air is lighter and rises. The teacher discusses the relationship between cold air and high pressure and warm air and low pressure.

Objective   Locating sources of cold and warm air masses.

Teacher Preparation   Review the lessons on the earth's tilt and its revolution around the sun. Have a world globe and study lamp (or flashlight) available.

Activities   In a darkened room, students use the light as the sun and revolve the tilted earth around the sun. They observe that the north and south poles receive sunlight at very low angles and that sometimes the poles receive no sunlight at all. They observe that the area near the equator receives very direct sunlight year round. From their previous lessons, they conclude that the polar regions are the source of very cold air and the tropical zone is the source of very warm air. The teacher locates the small arrows on the globe, which indicate the movement of large air masses from the polar and tropical regions.

Investigating Water in the Air

General Overview The cyclic heating and cooling of the earth causes water to constantly move into and out of the atmosphere. Recognizing and understanding the processes of evaporation, condensation, and precipitation will enable students to comprehend many common occurrences in their surroundings.

Learning Objectives Students will identify some factors affecting evaporation and condensation of water. They will also learn to recognize some basic cloud types and forms of precipitation.

MON

Objective Observing the effect of radiant heat on rate of water evaporation into the air.

Teacher Preparation Provide each pair of students with two aluminum pie pans and thermometers. Have some measuring spoons available or mark a position on a glass for measuring purposes.

Activities Using measuring spoons, students pour equally small amounts of water into two aluminum pie pans. They place one pan on the window sill in sunlight or under an incandescent light bulb. They place the other pan of water in a darker area of the room. They place a thermometer in each pan and observe differences in temperature. They observe the rate of evaporation in the two pans and conclude that radiant heat causes water molecules to evaporate into the air faster. They discuss how sunlight affects the water in rivers, lakes, and oceans.

TUES

Objective Describing how humidity and wind affect evaporation of water.

Teacher Preparation Give small groups of students three aluminum pie pans and one clear plastic bag large enough to fit over one of the pie pans. Set up an electric fan at one end of the room.

Activities Students pour equal amounts of water into three pie pans and place the three pans in a sunny location. They place a plastic bag over one pie pan, place the second pan in front of a blowing fan, and keep the third pan as a "control." They predict which water will evaporate the fastest and then observe what happens. After several hours, they place their hand inside the plastic bag and feel the moisture. The teacher describes the inside of the bag as being *humid*—the air in the bag contains more water vapor than the air outside. They discuss the difference between high and low humidity. From their results, they describe a set of conditions that would cause the greatest and least amount of evaporation.

WED

Objective Condensing water vapor from the air.

Teacher Preparation Make some ice cubes with colored water. Have available some clean tin cans, stripped of their wrappings, and some aluminum foil.

Activities Students place some colored ice cubes inside the tin cans and cover the tops of the cans with aluminum foil. They observe the moisture from the air that collects (condenses) on the outside of the cans. (Since the water on the inside of the can is colored and the water on the outside is clear, the suggestion that the water somehow passes through the can is discounted.) They relate this activity to their previous knowledge that cold air contracts and conclude that cold air holds less water than warm air. The teacher identifies dew, frost, fog, and clouds as examples of *condensation*.

THURS

Objective Describing cloud types and their associated weather.

Teacher Preparation Describe clouds as water vapor that has condensed around tiny dust particles in the air. Have available pictures or drawings of the three basic types of clouds and describe them in turn: (1) *cirrus*—the highest clouds in the sky, thin and wispy, usually meaning good weather; (2) *cumulus*—large, fluffy clouds that usually appear in the afternoon and disappear after sunset, usually associated with fair weather, but certain types are thunderstorm clouds; (3) *stratus*—low clouds that cover the whole sky and blot out the sun, usually associated with rain or snow. Have pictures of the *variations* of these cloud types available, too.

Activities Students learn to identify the three basic cloud types in the sky and associate the kind of weather with the types of clouds. They practice their cloud identification skills throughout the school year.

FRI

Objective Describing and recording types of precipitation.

Teacher Preparation Have some posterboard available. Define precipitation as a form of moisture that falls from the air. Show some pictures suggesting different forms of precipitation and ask the students to describe the different kinds of precipitation.

Activities Students identify rain, drizzle, sleet, snow, and hail as types of precipitation. They make a poster with the months and days of the school year on it. Then they make up some symbols for each form of precipitation and take turns recording the appropriate symbols on the days that precipitation occurs.

Notes

General Overview A study of weather affords the teacher a good opportunity to involve the students in using the past and present to project future events. By observing and collecting weather data, students can begin to understand patterns in nature. The devices that trained weather officials use in recording weather data are complex and exact, but young students can build crude weather devices that will yield sufficient information to report the weather with some degree of consistent accuracy.

Learning Objectives Students will learn to locate weather information from different sources. They will also learn how to gather and record certain weather information.

MON

Objective Locating sources of weather reports and studying weather information.

Teacher Preparation Read the weather forecast for the day and several other significant days from the *Farmer's Almanac* and discuss with the students how such long-range predictions are made. Have available several copies of weather reports in old newspapers.

Activities Students observe that newspaper weather reports usually have information about the temperature, precipitation, whether it will be sunny or cloudy, humidity, direction and speed of winds, and barometric pressure. They listen to the weather report on radio or television several times during the school day and observe that most of the same information is given. They discuss how they might build a weather center and gather some of the same information at school.

TUES

Objective Measuring and recording temperature changes.

Teacher Preparation Provide an outdoor thermometer and a posterboard chart for recording temperatures.

Activities Students take turns recording the temperature on a class chart at different times every day throughout the school year. They circle the low and high temperatures at the end of each day. They discuss patterns in their data: lows in the mornings, highs in the afternoons; gradually higher or lower temperatures as the seasons change; relationship to cloudy or sunny days.

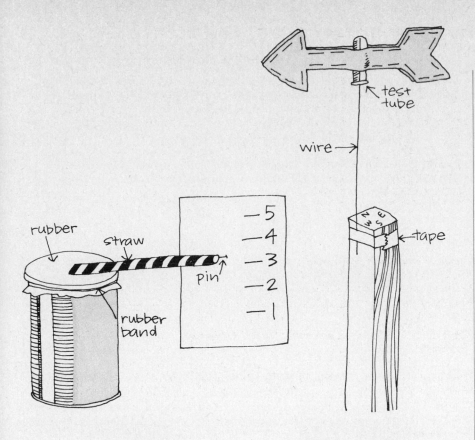

Objective   Making and recording wind direction.

Teacher Preparation   Provide several aluminum pie pans, wire clothes hangers, and test tubes. Have available some scissors, wire cutters, masking tape, and compasses. Locate some positions in the school grounds to put the weather vanes.

Activities   Students cut two identical arrow shapes from large aluminum pie pans. They staple the two arrows together along the edges of the head and tail. They find the balancing point of the arrow, insert a small test tube at that point between the two pieces of aluminum, and staple the edges of the arrow shaft. With wire cutters, they cut a straight piece of wire from a coat hanger and fasten the wire to a wooden pole stuck in the ground. They use a compass to mark north, south, east, and west on the pole. They observe the arrow pointing toward the direction from which the wind is blowing. They record the wind direction on a daily basis. They look for a relationship between wind direction and type of weather.

Notes

WED

Objective   Measuring changes in air pressure.

Teacher Preparation   Solicit student help in bringing some large tin cans to class with their tops removed. Cut up an old rubber inner tube or balloons into sizes large enough to wrap tightly over the tops of the cans. Have available some rubber bands, soda straws, straight pins, cardboard, blocks of wood, glue, and rulers. Discuss the fact that air exerts pressure and that the amount of pressure changes. Identify the *barometer* as the device that measures changes in air pressure.

Activities   Students use some string or rubber bands to secure the rubber to the can. They glue a pin to one end of a soda straw, and they flatten the other end and glue it to the rubber top. They make a cardboard scale with marks spaced at equal distances and record daily rises and drops in pressure as indicated on the scale. Students compare what happens with their barometers with what the weather reports say about the *barometric* readings.

THURS

Objective   Collecting and measuring precipitation.

Teacher Preparation   Have available several different containers: large and small tin cans, large-mouth glass jars, and milk cartons with the top half cut off. Solicit student help in making a classroom chart of precipitation.

Activities   Students use the various containers to collect some rain water. They measure the amount (depth) of water in each container and discover that the depth of water in each one is about the same. (The volume, of course, is different in larger and smaller containers.) The students take turns measuring the amount of precipitation that falls every twenty-four hours during school days and recording the data on a classroom chart. They compare their results with those provided by the newspaper, radio, and television weather reports.

Identifying Factors That Weather Rock

General Overview Except in the rare occurrences of earthquakes, volcanic eruptions, and human-made explosions, the earth's surface doesn't appear to change very much. However, daily changes in weather affect the earth's surface constantly. Weathering of rock produces both soil and sandy beaches. The wearing down of rock is highlighted in the twenty-second week.

Learning Objectives Students will observe that soil particles come from larger rocks. They will investigate some of the ways in which rocks are reduced to smaller particles.

MON

Objective Investigating the hardness and other characteristics of rocks.

Teacher Preparation Solicit students' help in bringing a variety of different rocks—hand size and smaller—to class. Have a hammer available.

Activities Students rub different kinds of rocks together over sheets of paper. They observe the texture and appearance of the particles resulting from the rubbing action and compare the particles to the rock that produced them. They also observe that some rocks are harder than others and produce less dust when rubbed. They try to crush some of the small particles that are broken off. They discuss and compare their observations of hardness, color, and texture of the larger rocks and the smaller particles produced from the larger rocks. The teacher summarizes the results in chart form, emphasizing the difference in wear that rubbing has on the different rocks.

TUES

Objective Observing the effect that moving water has on rocks.

Teacher Preparation Provide some soft rocks that break easily such as sandstone, limestone, and shale. Also have a hammer, thick glass jar with a lid, plus another jar available.

Activities Students observe as the teacher breaks a larger rock into small pieces; they see and feel the rough edges. They observe as the teacher puts some rock pieces into two jars, fills them about half full with water, and screws the lids on tightly. One jar of rocks and water is left on the teacher's desk. Students take turns shaking the second jar about twenty times each. When the jar has passed around the room, they compare the clarity of water in both jars and see and feel the rocks from both jars.

WED

Objective Observing the effect that freezing temperatures have on rock.

Teacher Preparation Soak some limestone, sandstone, shale, or brick in water overnight. Put the rock in a plastic bag and place it in the cafeteria freezer for several hours.

Activities Students observe that some of the rock breaks off as a result of the freezing water expanding inside the rock. They discuss what areas of the earth might be affected in the same way—any area that experiences freezing temperatures.

THURS

Objective Investigating the effect that changing temperatures have on rock.

Teacher Preparation Put some limestone, sandstone, or shale rock in an aluminum pie pan. Alternately heat and cool the rocks in the cafeteria oven and freezer every hour throughout the day.

Activities Students observe the effect that alternate periods of heating and cooling have on the rocks. They observe the smaller rock particles that break off from the larger rocks. They infer from their results what happens to rocks in deserts and on mountains due to warm days and cold nights.

FRI

Objective Observing that growing things can break up rocks.

Teacher Preparation Locate some areas around the school where the roots of trees have broken up the pavement, concrete sidewalks, or boulders. Show students some pictures of trees growing in cracks in rocks.

Activities Students observe the effect of the force of growing trees on pavement, concrete, and rocks and conclude that growing things are one cause of rocks breaking into smaller pieces. They summarize all the different factors that they have observed that can cause rocks to break up. All of the factors studied (and others, such as wind) are identified as examples of how rock is slowly worn down to soil.

Notes

General Overview An irreplaceable and very valuable natural resource is the earth's soil. The two major factors that affect soil erosion—water and wind—are the subject of the lessons in week 23.

Learning Objectives Students will identify the process of erosion and will recognize moving water and wind as two causes of erosion. They will also investigate some factors that affect the rate and amount of erosion.

MON

Objective Identifying the process of water erosion.

Teacher Preparation For each pair of students, provide an aluminum pie pan and a water sprinkler (you can use salt shakers or make your own sprinklers by punching holes in jar lids). Have about three pounds of dirt available. Cover an area on the floor or table with paper towels.

Activities Working in pairs, students build a small dirt "mountain" in their pie plates—not filling up the entire bottom area of the pan. They use their sprinklers to simulate rain on their mountains. They observe that the soil on top of the mountain is washed down toward the bottom, reducing the height of their mountain. They identify gullies and *deltas* where the soil is deposited at the bottom of the mountain. The teacher identifies this in nature as water *erosion*—wearing down the land.

TUES

Objective Comparing water erosion on different slopes.

Teacher Preparation Construct two stream tables by cutting the top off a cardboard box at least a half-meter (about ½ yd) long and as narrow as can be found. Cut the sides down to about 2.5 cm (1 in.) above the bottom. Cover the inside bottom and sides with plastic wrap or aluminum foil. Have two pans, some soil, and water sprinklers available.

Activities Students pack the two boxes full of soil and prop the boxes up at different angles. They observe that when both boxes of soil are given the same amount of water, the box with the steeper slope loses more soil and more quickly than the box set at a lower slope. They discuss locations such as mountains that might have greater erosion because of their steep slopes.

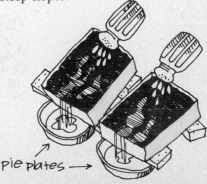

pie plates →

41

WED

Objective Observing the effects of grass on erosion.

Teacher Preparation Set up the same stream-table apparatus as in the previous lesson. Have some sod with grass growing on it and a bucket of soil.

Activities Students pack one box with soil; in the second box, they lay the grassy soil in the box so that it is level with the sides. They prop the boxes up at the same steep slope. When equal amounts of water are poured over both boxes, the soil covered with grass washes away slower and in smaller amounts. They conclude that grass reduces the rate of water erosion.

THURS

Objective Investigating wind erosion.

Teacher Preparation Provide a window or oscillating fan, 2 liters (2 quarts) of dry sand, and a bough from a pine tree (or any other tree).

Activities Students assist the teacher in placing two piles of dry sand on separate pieces of posterboard. The fan is allowed to blow directly on one pile of sand for about 10 minutes. Students observe the movement of the sand particles. The fan is then turned to blow from the same distance on the second pile of sand, but a pine bough is placed between the fan and the sand. Students compare the movement of the sand and conclude that wind can cause erosion, but that trees can reduce the rate and amount of erosion.

FRI

Objective Observing erosion on the school ground.

Teacher Preparation Locate several sites around the school where erosion is occurring.

Activities Students look for areas where no grass is growing and where water runs often—at downspouts, beneath window air conditioners, in gullies, and behind the cafeteria. They also locate areas that are exposed to open wind. They return to the same locations often to observe changes. They suggest ways of reducing erosion around the school.

Notes

24

MON

Objective Identifying and comparing plants that produce seeds from flowers.

Teacher Preparation Have available films and reference books with pictures of the different types of flowering plants, including trees, grasses, shrubs, vines, vegetable plants, water plants, and ornamental flowers. Lead a brief discussion describing one major category of plants as those that produce seeds from flowers. Ask students to examine the reference materials for descriptions of the different types of flowering plants. Take them on a field trip around the school and into a wooded area if possible.

Activities Students identify some of the many different types of flowering plants from the reference materials and on the field trip. They observe the different plants throughout the school year, noting when different plants are flowering. They compare differences in size, texture, odor, and shape of the plants, including their stems, leaves, and flowers.

General Overview Green plants are plants that convert their own nutrient needs from sunlight, air, and water. Learning to identify the different categories of green plants and their environments is important beyond the need to be generally knowledgeable about plants. It contributes to a broader understanding of the interrelationships of plants and animals in different living communities. Students should not be expected, however, to memorize names and categories of plants.

Learning Objectives Students will identify five major groups of plants: flowering plants, cone-bearing plants, ferns, mosses, and algae.

TUES

Objective Identifying plants that produce seeds from cones.

Teacher Preparation Have available films and reference books with pictures of different types of cone-bearing plants: pines, spruces, firs, hemlocks, redwoods, and cedars. Lead a brief discussion concerning the category of plants that produce seeds in cones, often called "evergreens." Ask students to describe some of the different types of cone-bearing plants. Take them on a field trip to locate some cone-bearing plants and collect some young, closed cones with seeds and some old, open cones without seeds.

Activities Students identify cone-bearing plants in the reference materials and on the field trip. They examine them, locating seeds in the young cones and observing how the old cones have opened to release the seeds. They compare and contrast the size, texture, odor, and shape of the cone-bearing and flowering trees. They especially notice differences in the bark and leaves.

WED

Objective Identifying spore-producing plants: ferns.

Teacher Preparation Have available films and reference books with pictures of different types of ferns, including the large tropical fern trees. Lead a brief discussion identifying ferns as a category of plants that reproduce by spores, not seeds. Take them on a field trip to locate some ferns around the school, especially in wooded areas. Bring some potted ferns to class. Provide the students with magnifiers.

Activities Students identify different types of ferns in the reference materials and on a field trip. They observe the close relationship between ferns and moist surroundings. Using the magnifiers, they observe the spore cases on the underside of the fern leaflets.

THURS

Objective Identifying other spore-producing plants: mosses.

Teacher Preparation Have available films and reference books with pictures of mosses. Identify mosses as another category of green plants that reproduce with spores. Take students on a field trip to locate mosses growing around the school. Provide the students with magnifiers.

Activities Students learn to identify mosses (though probably not different types) in the reference books and on the field trip. They observe that mosses are found more often in shady and moist locations. They observe the stalklike spore cases of mosses with their magnifier.

FRI

Objective Identifying and describing algae.

Teacher Preparation Have available films and reference books with pictures of saltwater, freshwater, and land algae. If possible, take children on a field trip to a pond to collect algae. Schools near the ocean could conduct a "seaweed"-collecting field trip.

Activities Students identify the different types of algae in the reference materials and on the field trip. They discover that the terms "seaweed" and water "scum" usually refer to algae. They also discover that not all algae are green.

Notes

Studying How Seed Plants Reproduce

General Overview   One of the basic concepts of life science is that living things reproduce. Sexual reproduction is the major form of reproduction in plants (as well as animals), but plants can also reproduce vegetatively. The activities of this week emphasize related aspects of plant reproduction.

Learning Objectives   Students will identify the role of flowers in plant reproduction as well as how they are pollinated. They will also learn about some of the food by-products of plant fertilization and how seeds are dispersed. Finally, students will learn how plants reproduce other than by seeds.

MON

Objective   Identifying the role of flowers in plant reproduction.

Teacher Preparation   Provide some large flowers and magnifiers for each student and a simple drawing of the flower parts. Explain that when pollen is carried from the stamens to the pistil of the same flower or another flower, it is called *pollination*; that when the pollen grain joins with the "eggs" in the ovary, it is called *fertilization*; that seeds develop from fertilized eggs; and that the ovary ripens into a fleshy or dry fruit. Take students on a field trip to locate different kinds of flowers, fruits, and seeds.

Activities   Students use the flower drawing to identify the parts of a flower. Using the magnifiers, they observe pollen grains on the stamens of a flower. They use their fingernails to open up the ovary and to view the eggs.

TUES

Objective   Describing how plants are pollinated in nature.

Teacher Preparation   Lead a brief discussion on how plants are pollinated by insects and wind. Use visuals to supplement the discussion. Take students on a field trip to observe bees, wasps, and other insects pollinating flowers. Show them the accumulation of pollen on window sills and car windows that blows off pine and oak trees in the spring. Point out the large amount of pollen produced by plants. Ask the students to suggest the advantage of such large amounts of pollen for plant reproduction.

Activities   Students observe evidence of pollination by insects and wind. They suggest that large amounts of pollen make it more likely that the female eggs will be fertilized.

WED

Objective   Identifying different types of fruit.

Teacher Preparation   Ask the students to name some fruits they are familiar with—most responses will probably refer only to products of tree fruits, such as apples, pears, and bananas. Identify some other fruits: cereal grains such as oats and wheat; nuts such as acorns and walnuts; "vegetables" such as tomatoes, peas, and squash; and berries such as blackberries and raspberries. Use visuals and actual fruits to supplement the discussions.

Activities   Students suggest additional types of fruits and categorize them according to types. The teacher lists their suggestions on the board.

THURS

Objective   Identifying how seeds spread to different areas.

Teacher Preparation   Have available films and reference books with pictures showing different methods and examples of seed dispersal: wind (dandelion); wings (maple), stickers and animal fur (sandspur), and birds (virtually any small seed). Lead a discussion and field trip around the school to identify different types of seed dispersal.

Activities   Students identify various ways that seeds are spread to different locations.

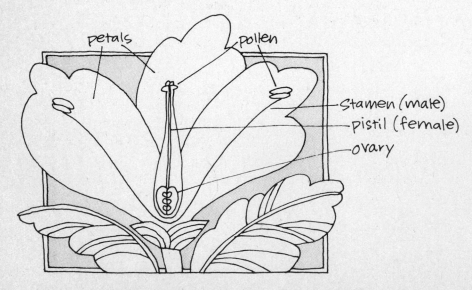

FRI

Objective Observing plants that reproduce without fertilization.

Teacher Preparation Lead a discussion on how plants can sometimes reproduce without seeds. Show the students pictures of plants reproducing by tubers (Irish potato), runners (strawberries), roots, stems, and leaves. Provide different groups of students with materials to grow plants from roots, stems, and bulbs: sweet potatoes (a root); cutting of coleus, English ivy, or wandering jew; and narcissus bulbs.

Activities Students suspend the sweet potatoes in water. They place the stem cuttings in water and transfer them to potting soil after roots are established. They place the large end of the narcissus bulbs down into a low saucer filled with pebbles and water. They observe the growth over a period of weeks and months.

Notes

General Overview In studying how plants reproduce, students need to become aware of seeds as products of green plants and some factors associated with their germination. The teacher should begin all the activities calling for germinating seeds on Monday or the previous Friday to allow for observation of results throughout the week. The activities of this week will probably require more than five days to complete.

Learning Objectives Students will identify seeds as products of green plants, observe the internal structure of seeds, and investigate some factors relating to seed germination.

MON

Objective Identifying seeds as products of green plants.

Teacher Preparation Discuss with students the fact that green plants produce seeds and germinating seeds produce new plants. Have available a large variety of seeds such as unshelled peas, beans, and nuts, plus a variety of other seeds. Have some resource books available showing pictures of seeds.

Activities Students observe the variety of seeds and associate them with the plants that produced them.

TUES

Objective Investigating the inside of a seed.

Teacher Preparation Provide a magnifier and some bean seeds that have been thoroughly washed (to remove chemicals that retard germination). Provide a variety of other large seeds such as shelled peas and peanuts.

Activities Students observe the inside of the bean seed, identifying the embryo plant and the food supply. They locate the embryo and food supply in other seeds.

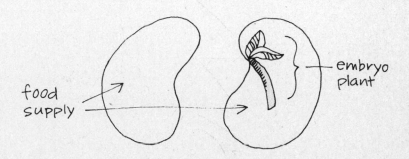

food supply — embryo plant

WED

Objective Investigating the need for soil and water for seed germination.

Teacher Preparation Provide each group of students with some radish, grass, or bird seed, two flat sponges, and two bowls. Have them place each sponge into a separate bowl and pour water into one bowl. Have them place some seeds on top of both sponges.

Activities After a few days students observe that seeds placed on top of the wet sponge germinate, while the seeds on the dry sponge do not germinate. They conclude that seeds need water, but not soil, to germinate.

paper towel

seed

THURS

Objective Investigating the need for light and warmth for seed germination.

Teacher Preparation Provide each group of students with radish, grass, or bird seed, two flat sponges, and two bowls. Have them place a sponge in each bowl and pour water into both bowls. Have them place some seeds on top of both sponges. Have a refrigerator available that has the inside light screwed loose so as not to turn on when the door is opened.

Activities Students place one bowl in a refrigerator and one in a dark, warm place in the room. In a few days they observe that few, if any, seeds placed in the refrigerator have germinated, while most, if not all, seeds placed in the other bowl germinated. They conclude that seeds need warmth, but not light, to germinate.

FRI

Objective Investigating the growth direction of roots and stems in seedlings.

Teacher Preparation Provide each student with a clear plastic tumbler, paper towels, and seeds. Have them wrap wet paper towels around the inside walls of the tumbler. Have them place seeds between the paper towels and tumbler at different orientations. Have them moisten the towels daily.

Activities Students observe that, no matter how the seeds are oriented in the tumbler, the roots always grow down and the stems grow up. They discuss the advantages of this growth pattern for growing seedlings.

Investigating Life Needs of Green Plants

General Overview Most all living things have some common necessities for life: water, air, warmth, and food. How animals and green plants obtain food is very different. Whereas animals have to capture their food, green plants can produce their own food, if the other life necessities are present. The activities of this week are concerned with identifying some of the life necessities of plants, leading up to an understanding that plants produce their own food by a process known as *photosynthesis*. The suggested activities for this week will require more than five days to complete.

Learning Objectives Students will identify light, water, and chemicals from soil and air as necessities for plants to carry on the food-making process (photosynthesis). They will also observe how water and chemicals from the soil move upward in plants.

MON

Objective Investigating the necessity of light for green plants.

Teacher Preparation   Provide pairs of students with growing seedlings planted in styrofoam cups. Have them place one cup in a dark place at room temperature and water regularly. Ask them to leave the other cup of seedlings in the lighted room and water regularly. Provide rulers.

Activities Students note the green color of the leaves, the number of leaves, and the length of stems and leaves of the seedlings in both cups on the day they begin investigation and every other day thereafter until significant changes are observed or until the plants left in the dark appear dead. Students conclude that sunlight is necessary for green plants to live.

TUES

Objective Investigating the necessity of water for green plants.

Teacher Preparation   Provide small groups of students with growing seedlings planted in styrofoam cups. Ask them to water one cup of seedlings on a regular schedule, every three or four days. Ask them not to water the seedlings in the second cup at all.

Activities Students observe that plants receiving water continue to live and grow, while plants not receiving water wilt, turn yellowish-brown, and die. They conclude that water is necessary for plants to survive.

WED

Objective Investigating the necessity of soil minerals for green plants.

Teacher Preparation Provide each pair of students with a sponge, bowl, styrofoam cup, potting soil, and seeds. Have students plant some seeds in moist potting soil in the cup. Also have them place some seeds on top of a wet sponge in a bowl. Be sure that both the potting soil and sponge are kept moist. Provide some rulers.

Activities   Students observe that seeds germinating in the soil continue to grow and develop new leaves, while seeds that germinate on the sponge eventually stop growing and die. They conclude that the soil somehow provides the seedlings with some chemicals necessary to make food. (If students suggest that plants get their food from the soil, ask them why in the Monday activity the seedlings growing in the soil and placed in the dark did not live and grow like those in light.)

THURS

Objective Observing that water and other chemicals move upward in plants.

Teacher Preparation Cut the bottom off some fresh celery stalks and place the cut ends in some water containing red food coloring. Have some magnifiers available. Following the students' observations that water moves up in plants, explain to the students that chemicals in the soil are mixed with the water (like the food coloring) as it moves up to the leaves.

Activities Students observe the colored water slowly rising in the celery stalks until the tops turn red. They use their magnifiers to observe the celery cross sections and colored "threads" which they pull away from the stalk. They conclude that water-tube cells in the celery stalk carry (transport) the water and chemicals from the soil up to the leaves.

28

General Overview Not all plants are green. The nongreen plants, called *fungi*, are a group of plantlike organisms that obtain their nourishment from living and dead things. As a group, fungi are both very beneficial and harmful to human activity. It is important that students have some minimal knowledge of the different types of fungi and their effects on humans.

Learning Objectives Students will identify and investigate mushrooms, puffballs, lichens, bread mold, mildew, rusts, smuts, and yeast cells.

FRI

Objective Identifying photosynthesis as the process of making food in green plants.

Teacher Preparation Lead a discussion summarizing what plants need to make food—sunlight, water, soil, and air—but that a chemical in the air called *carbon dioxide* moves into the leaves and stems of green plants and is used in making the food. Identify the food-making process in green plants as *photosynthesis*. Describe one very important chemical that moves out of plant leaves as *oxygen*—a chemical that humans and other animals need for survival.

Activities Students participate in the discussion.

MON

Objective Collecting and observing mushrooms, puffballs, and lichens.

Teacher Preparation Display reference books with pictures of mushrooms, puffballs, and lichens. Discuss the use of cultured mushrooms as food and the danger of eating wild mushrooms. Tell the students that these and all other nongreen plants obtain their nourishment from decaying matter or from other living things. Lead the students on a field trip around the school grounds and neighboring wooded areas to collect some mushrooms and puffballs.

Activities Students collect and observe different kinds of mushrooms and puffballs. They place them on the window sill and/or near a heat vent to dry them out. They group them according to similarities and differences. They use reference books to determine their names.

TUES

Objective Growing bread mold.

Teacher Preparation Provide each student with a small portion of sliced bread. Ask the students to wipe their pieces of bread across a dusty surface and then moisten the bread. Have them place the bread in a warm, dark location. Have magnifiers available.

Activities In a few days students observe mold growing on the bread. They observe it closely with the magnifiers, making drawings of what they see. They discuss where the mold obtains its nourishment for growth. They discuss other foods that become moldy if not protected from moisture and warmth.

WED

Objective   Locating mildew.

Teacher Preparation   Identify mildew as a type of fungus that grows on clothes, books, and other plants, especially in humid conditions. Show students pictures of mildew or actual articles that have mildew growing on them. Have magnifiers available. Discuss where and when mildew is likely to grow at home—in closets, bathrooms, and basements; and how it might be prevented—proper ventilation, airing shoes and clothes before putting away, and cleaning with disinfectants.

Activities   Students observe the picture of mildew and observe mildew growth with their magnifiers. They identify areas of mildew growth at home and attempt to prevent future mildew growth.

THURS

Objective   Identifying rusts and smuts.

Teacher Preparation   Have reference books and pictures about corn smut, Dutch elm disease, pine blister rusts, and apple rust. Collect some leaves showing rusts (most elms and apple trees exhibit some infection). Have magnifiers available.

Activities   Students read and examine pictures about rusts and smuts and their effects on trees and food crops. They examine some infected leaves with magnifiers.

FRI

Objective   Observing the growth of yeast cells.

Teacher Preparation   Display some books with pictures of yeast cells and some of their benefits to human activity. Identify yeast as a type of fungus. Pour two glasses full of warm water. Dissolve one teaspoon of sugar in one of the glasses. Add one package of dry yeast to both glasses.

Activities   After about twenty-four hours students observe that the glass containing the sugar shows evidence of a chemical change, but the glass without the sugar shows no change. From the reference books and classroom discussion, they conclude that the yeast cells are growing and reproducing and that the food for this came from the sugar.

Notes

Identifying Animals without Backbones

General Overview There are, of course, large numbers of animals known to elementary children. However, students do not necessarily associate animals as belonging to larger groups of related animals. Though we discourage involving elementary students in memorizing scientific classifications of animals, we encourage tasks that involve associating related animals. A major division of animals should be recognized by students: animals with backbones, called *vertebrates*, and animals without backbones, called *invertebrates*. The activities of week 29 are concerned with associating groups of animals without backbones.

Learning Objectives Students will identify classes of invertebrate animals identified as simple animals, worms, spiny animals, mollusks, and animals with outer skeletons.

MON

Objective Identifying and drawing the simple animals.

Teacher Preparation Have available films and reference books with pictures of sponges, jellyfish, sea anemone, and coral. Make a posterboard chart labeled "Simple Animals." Ask the students to draw and color some examples of the simple animals and to determine where they live (their *habitat*).

Activities Students label their drawings with names given in the reference books and display them on the chart. They discuss where these animals live and share their knowledge of, and past experiences with, some of the simple animals.

TUES

Objective Identifying and drawing the different categories of worms.

Teacher Preparation Have available films and reference books with pictures of: (1) flatworms, (2) roundworms, and (3) earthworms and leeches. Make a posterboard chart labeled "Worms" with subheadings for each of the three major types of worms. Ask the students to draw some examples of the different types of worms and locate information about where they live.

Activities Students label their drawings with the names given in the reference books and display them in the proper category on the chart. They discuss where these animals live and some of the beneficial and harmful effects of the worms on human activity—for example, the parasite flatworms and roundworms that infect animals, including humans, and plants.

WED

Objective Identifying and drawing the spiny animals.

Teacher Preparation Have available films and reference books with pictures of starfish, brittle stars, sea urchins, sand dollars, and sea cucumbers. Make a posterboard chart labeled "Spiny Animals." Ask the students to draw some examples of the spiny animals and locate information about where they live. Also encourage any students who have them to bring their sea-urchin and sand-dollar collections to class.

Activities Students label their drawings with the names given in the reference books and display them on the chart. They observe the radial symmetry displayed by the animals. They discover that all the spiny animals live in the oceans.

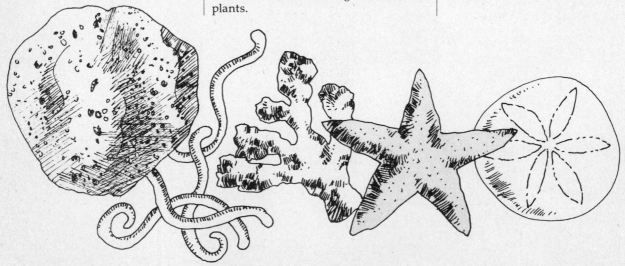

THURS

FRI

Objective Identifying mollusks.

Teacher Preparation Have available films and reference books with pictures of land snails and slugs, plus the ocean mollusks: clams, oysters, scallops, mussels, conches, whelks, periwinkles, limpets, abalones, squids, and octopi. Set up a "Mollusk Center" in the room. Encourage students to take part in different activity options after they have examined the reference books on mollusks, looking for similarities and differences in mollusks.

Activities Students participate in one or more activities: bring shell collections to school and identify the types of shells; make colored drawings of different mollusks to display in the center; collect some snails and slugs from home and observe their behavior in a snail terrarium that they construct out of empty milk cartons; observe and describe the behavior of freshwater snails in the classroom aquarium; or report on the foods derived from mollusks.

Objective Identifying animals with outer skeletons.

Teacher Preparation Have available films and reference books with pictures of shrimps, crabs, lobsters, crayfish, insects, centipedes, millipedes, spiders, and ticks, all of which are examples of a group of animals called *arthropods*. Set up a center for learning about "Animals with Outer Skeletons." Encourage students to take part in different activity options after they have examined the reference books on arthropods, looking for similarities in this group of animals.

Activities Students participate in one or more activities: make an insect collection and identify the specimens; observe and describe the behavior of ants in an "ant farm"; grow some butterflies or moths from the cocoon stage; report on bee keeping; report on the helpful and/or harmful effects of different insects; report on the food value of ocean arthropods; describe the difference between insects and spiders; report on poisonous and nonpoisonous spiders; or draw different kinds of insect and spider "homes."

Identifying Animals with Backbones

General Overview As previously mentioned, all animals can be classified into two major groups: those that have backbones and those that do not. Animals with backbones form the group of larger animals to which humans belong. The different classes of vertebrates are rather easily identified and learned by students. Students can also discriminate major subgroups within the classes of vertebrates. Knowledge of the classes and subgroups will broaden the students' understanding of diversity of form and function in nature. Throughout the week, encourage the students to bring the bones from different animals to class in order to compare skeletal features.

Learning Objectives Students will identify and describe classes and groups of vertebrate animals: fishes, amphibians, reptiles, birds, and mammals.

MON

Objective Identifying and describing different fishes.

Teacher Preparation Have available films and reference books with pictures of various freshwater and saltwater fish and eels. Ask the students to describe some of the different kinds of fish with which they are familiar. Lead a discussion on some of the similarities of fish they have seen and caught to others seen and described in the reference materials. Ask students to bring some clean fish bones to school after their next fish dinner at home.

Activities Students discover that all fish have gills, fins, and scales, lay eggs without shells in water, and are cold-blooded—the body temperature changes with the outside temperature. They examine and compare the variety of fish bones brought to school, observing the backbone vertebrae in particular.

TUES

Objective Identifying and describing different amphibians.

Teacher Preparation Have available films and reference books with pictures of different amphibians. Describe the two small groups of amphibians: those with tails, the salamanders; and those without tails, the toads and frogs. Take the students on a field trip to a stream, pond, or natural spring to look for some amphibians, their eggs, and some tadpoles.

Activities Students note that amphibians: produce their young from unshelled eggs laid in the water; live as adults on the land, usually in moist locations; and are cold-blooded. They list examples of different groups of amphibians and describe some of their differences.

WED

Objective Identifying and describing different reptiles.

Teacher Preparation Have available films and reference books with pictures of different reptiles. Identify lizards, snakes, turtles, alligators and crocodiles, and the extinct dinosaurs as groups of reptiles. Encourage students to take part in different activity options after they have examined the reference materials.

Activities Students discover that all reptiles have scaly skins, are cold-blooded, and that most produce their young from soft-shelled eggs. Students participate in one or more activities: bring different safe reptiles to school in shoe boxes; demonstrate gentle handling and care of reptiles; make a visual report on poisonous and nonpoisonous snakes; draw and label some of the extinct dinosaurs; or report on sea turtles and/or the giant land tortoises.

THURS

FRI

Objective Identifying and describing different birds.

Teacher Preparation Have available films and reference books with pictures of different birds. Ask students to identify some similarities and differences in birds. Also ask students to list examples of different groups of birds: gulls, water birds, birds of prey, perching birds, ground birds, and woodpeckers.

Activities Students discover that all birds have feathers, hatch their young from hard-shelled eggs, and are warm-blooded—the body temperature stays about the same regardless of the surrounding temperature. They list examples of the different groups of birds. They also list some of the differences in habitat, beaks, feet, and feeding habits.

Objective Identifying and describing different mammals.

Teacher Preparation Have available films and reference books with pictures of different mammals. Ask students to identify some similarities and differences in mammals. Also ask students to list examples of different groups of mammals: those that fly, live in oceans, eat insects, have hooves, have pouches, gnaw, and walk upright.

Activities Students discover that all mammals are warm-blooded, have some hair, produce live young, and that female mammals *nurse* their young with milk. They list examples of the different groups of mammals.

Observing Animal Needs and Reproduction

General Overview An important concept for students to comprehend is that even though animals differ in structure, they have some common life-sustaining needs: food, water, and air. Another important concept to develop in the elementary years is that living things, including animals, reproduce offspring similar to themselves. Each activity in the thirty-first week will emphasize the life needs and reproduction of different animals, including the students' pets at home. Many of the activities of this week could be followed throughout the school year.

Learning Objectives Students will observe and describe the life needs and reproduction of: classroom fish, mammals, and reptiles; schoolyard birds; and home pets.

MON

Objective Observing and describing the life needs and reproduction of aquarium fish.

Teacher Preparation Solicit student participation in setting up a balanced classroom aquarium with living plants, snails, and male and female guppies. Encourage students to care for the aquarium and to observe the movements, reproduction, and growth of the guppies. Have magnifiers, guppy food, reference books, and pictures of fish available.

Activities Students count the number of guppies and observe any changes in size and number. They observe the movement of the guppies with and without the aid of a magnifier. They especially observe the motion of the mouth and gills and discuss how the guppies get air. Students take turns feeding the fish. From students' past experience with fishing, they describe what happens when fish are caught and left out of the water. In a discussion, they conclude that the fish need food, water, and air, and that they move, grow, and can reproduce fish like themselves.

TUES

Objective Observing and describing the life needs and reproduction of classroom mammals.

Teacher Preparation Provide a suitable cage for some male and female guinea pigs or white rats. Encourage students to care for the mammals and cage and to observe the movements, reproduction, and growth of the animals. Have a tape measure and a set of scales available to measure and weigh the animals.

Activities Students take turns feeding and watering the guinea pigs, as well as cleaning their cage. They measure and weigh the guinea pigs once a week and record the changes. They observe their breathing and their motions throughout the day. They also observe the birth of baby guinea pigs. Students conclude that the classroom mammals need food, water, and air, and that they grow and can reproduce animals like themselves.

WED

Objective Observing and describing the life needs and reproduction of turtles.

Teacher Preparation Provide several small turtles from a pet store and an appropriate container. The container—an aquarium will do—should be partly filled with water and have rocks large enough for the turtles to climb onto, out of the water. Have a supply of commercial turtle food available. Encourage the students to take turns providing the turtles fresh food, such as very small bits of lettuce, hard-boiled eggs, and raw hamburger. Have some pictures of turtle eggs being deposited and hatching.

Activities Students take turns caring for the turtles and the container. They measure and weight the turtles every week or so until they are grown. They observe the air openings in the "nose." They observe the movement of the turtles in and out of the water. The teacher shows them pictures of turtles laying eggs and the eggs hatching baby turtles. Students conclude that turtles need food, water, and air, and that they grow and can reproduce turtles like themselves.

THURS

Objective Observing and describing the life needs and reproduction of birds.

Teacher Preparation Set up a bird feeder outside the window of the classroom or in some convenient location near the classroom. In a more remote location, perhaps in a wooded area adjacent to the school grounds, put up a bird house or two. Have some books showing pictures of different types of birds.

Activities Students use the reference books and learn to identify some birds common to their area, such as cardinals, robins, and blue jays. They locate the air openings in the beaks of birds. Students take turns placing bird seed and water in the feeder and observe the birds that feed throughout the year. They observe the bird houses occasionally to determine if any birds are nesting. If nests are established, they watch as eggs and newborn birds hatch and grow. They refer to pictures of bird eggs and newborn birds. Students conclude that birds need food, water, and air, and that they grow and reproduce birds like themselves.

FRI

Objective Observing and describing the life needs and reproduction of home pets.

Teacher Preparation Prepare space for a bulletin board on pets. Have colored construction paper, scissors, and paste available.

Activities Students bring pictures or make drawings of pets that they have at home. They paste their pictures and drawings onto construction paper. With the teacher's assistance they print or write the pet's name above its picture and some information about the pet below the name, such as age, number of offspring, what it eats, etc. After putting all the pictures on the bulletin board, students take turns telling the class about their pets.

From the week's activities, students conclude that all animals, no matter where they live and how different they look, need food, water, and air, and that they grow and can reproduce animals like themselves.

Notes

55

Identifying Survival Adaptations

General Overview Survival of the fittest is a key concept that refers to living organisms being favorably adapted to their surroundings. Organisms that can procure life's necessities, avoid being killed, adjust to changes in the physical environment, and reproduce their young are described as being adapted to their environment. Elementary students can identify examples of adaptations favorable to survival of living things even though they cannot yet comprehend the genetic factors operating. Students can also extend examples of adaptive survival to human activities. It is important during these lessons to avoid teleological references such as "giraffes developed long necks in order to get food from trees."

Learning Objectives Students will identify examples of plant and animal adaptations favorable to food gathering and protection in different surroundings. They will extend their examples to humans.

MON

Objective Identifying examples of adaptations for gathering or making food.

Teacher Preparation Have available films and reference books on plants and animals in different environments for use the entire week. Introduce the concept of plants and animals exhibiting adaptations that are favorable for making food (plants) and gathering food (animals and fungi). Provide some examples, such as a plant bending toward light, snake venom, and the long necks of giraffes. Ask students to suggest other examples. List these suggestions on the chalkboard and solicit student ideas for grouping the adaptations into categories, such as: speed, sensory ability, body structure (claws, fangs, beaks, teeth), trapping ability (webs, poison, stickiness).

Activities Students use reference materials to identify additional examples of adaptation for gathering or making food and list them according to categories.

TUES

Objective Identifying examples of adaptations that provide protection from being killed and eaten.

Teacher Preparation Introduce the concept of plants and animals exhibiting adaptations that provide protection from being eaten. Provide some examples, such as thorns on plants and the odor of skunks. Ask students to suggest other examples. List their ideas on the chalkboard and solicit student suggestions for grouping them into categories, such as: protective colors, hard and thorny surfaces, protective noises, poisons, sensory adaptations, size, and speed.

Activities Students use reference materials to identify additional examples of adaptations that provide protection from being eaten. They discover that some adaptations that are favorable for gathering or making food are also favorable for protection from being killed and eaten.

WED

Objective Identifying adaptations favorable to living in different surroundings.

Teacher Preparation Introduce the concept of plants and animals exhibiting adaptations that are favorable to different degrees of light, heat, water, air, and soil types. Identify those as the physical factors in our surroundings or physical *environment*. Provide some examples of living things adapted to the physical environment, such as gills in fish which obtain oxygen in water, blubber on whales which provides protection from the cold arctic waters, webbed feet in ducks, succulent plants in the desert, and lichens growing on bare rock. List additional adaptations identified by students on the chalkboard or posterboards.

Activities Students use references to identify additional examples of adaptations favorable to living in different surroundings.

THURS

Objective Identifying examples of adaptations that provide protection from changes in physical conditions.

Teacher Preparation Lead a brief discussion of living things exhibiting adaptations to changes in their physical environment, such as seasonal migration of birds and other animals, day lilies opening up in sunlight, bats coming out at night, waxes and oils on animal fur and skin which repel water, water-storage cells in cactus plants, seasonal changes in deciduous trees, and changes in animal coloration. List additional adaptations identified by students on the chalkboard or posterboard.

Activities Students use reference books to identify additional adaptations to changes in the physical environment.

FRI

Objective Identifying examples of how humans have changed their surroundings to provide food and protection.

Teacher Preparation Lead a brief discussion on human adaptations favorable both to obtaining food and to avoiding being eaten by other animals. Ask students to identify how different groups of people have modified their environment to ensure more efficiency in food gathering. Also ask them to identify how different groups of people exhibit adaptive characteristics favorable to different physical environments, such as black skin in tropical zones. Engage the students in a discussion of how different peoples have created ways to protect themselves from changes in the physical environments, such as heated and cooled buildings, electrical lights, and flood levees. List student suggestions on the chalkboard or posterboard for each area discussed.

Activities Students contribute ideas to the discussion.

Notes

General Overview An essential concept of life science for students to understand is the interdependency of plants and animals. Though the interactions of living things are much more complex than presented in these lessons, the basic ideas can be understood by elementary children. Lessons during this week emphasize food dependency in living things.

Learning Objectives Students will learn to identify food producers, consumers, and decomposers. They will also construct food chains, food webs, and number pyramids, and apply this knowledge to sources of human foods.

MON

Objective Identifying examples of organisms that are producers, consumers, and decomposers of food.

Teacher Preparation Using a picture of a nature scene showing different plants and animals, ask the students to identify the living organisms that produce food (plants) and those that consume food (animals). Ask them to further identify those animals that eat plants and those that eat other animals. Identify plants as *producers* and animals as *consumers*. Describe animals that eat plants as the *first consumers,* and animals that eat other animals that eat plants as the *second consumers.* Identify organisms that obtain their nourishment from dead organisms, like the fungi, as *decomposers*. Show students several different pictures of nature scenes to apply their new terms. Take a field trip around the school grounds to identify examples of the different categories of organisms.

Activities Students identify examples of producers, consumers, and decomposers in the nature scenes and on the field trip.

TUES

Objective Identifying food chains in nature.

Teacher Preparation Have the students write the name of an animal at the top of a sheet of paper. Under the animal, have them write one important food source for that animal. Identify the list as a food chain. Have them continue down the sheet until the producer organism is identified. Have them repeat this process for several other large and small animals, making long and short chains.

Activities Students list some food chains involving several consumer organisms. They identify green plants or products of green plants as the beginning of all food chains.

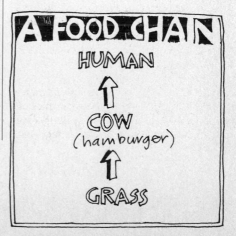

A FOOD CHAIN
HUMAN
↑
COW
(hamburger)
↑
GRASS

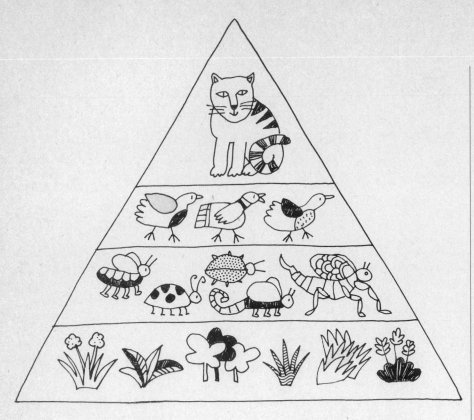

Objective Describing human foods and their sources.

Teacher Preparation Lead students in a discussion of the dependence of humans on plants and animals for all of their food. Have some old newspapers and magazines, scissors, and glue available. Make a posterboard with the heading "Human Food Sources."

Activities Students cut and paste on the posterboard examples of foods they eat and what plant or animal the food is derived from. After the collage is made, they identify examples in which humans act as first, second, or third consumers. They discuss the numbers of different types of organisms that humans eat as well as the amount of food humans eat.

Notes

WED

Objective Constructing a food web in the classroom.

Teacher Preparation Divide the class into two groups. Give each student in one group the name of a particular organism: flowering seed plant, bee, grub, mouse, raccoon, quail, king snake, owl, and fox. Station students in this group in different locations around the room. Provide students in the second group some string and scissors and assign them to stand beside one person in the first group. Then ask them to run the string from their organism to the one that it eats. For example, strings could be run from the snake to the mouse and to the grub. The teacher serves as the judge in any dispute over what organism eats what. Have reference books available.

Activities Students observe that the feeding interrelationship of animals is more complex than a simple food chain. They observe also that some animals can be eaten by each other—for example, adult owls eat young raccoons and adult raccoons can eat owl eggs.

THURS

Objective Observing a pyramid of numbers in the classroom aquarium and terrarium.

Teacher Preparation Lead a discussion of the amount of food it takes to keep the guppies or goldfish alive in the classroom aquarium. Discuss also the amount of food consumed by the turtles and lizards in the classroom terrarium. Relate these observations to the use of cats to control mice populations. Describe these as examples of *food pyramids*. Solicit additional examples of food pyramids from the students.

Activities Students participate in the discussion and describe other examples of food pyramids. They conclude that large numbers of producer organisms serve as food for a smaller number of the first consumers, which in turn become food for fewer numbers of the second consumers, and so on.

Studying Communities of Living Things

General Overview No living thing truly lives in isolation. All living things exist with others in groupings of plants and animals called *communities*. Differences in the availability of sunlight, heat, soil, water, and air result in different communities of living things. The activities of this week will introduce students to different communities of living things.

Learning Objectives Students will describe the interactions between organisms and their environment in different communities.

MON

Objective Describing populations of organisms in the schoolyard community.

Teacher Preparation Select some study sites on the schoolyard. Have student groups mark off a small area on the ground with a clothes hanger bent into a circle. Ask them to identify all the different types of plants and animals in the area. Let them assign arbitrary names to the organisms if they don't know the identity otherwise. Ask students to count the number of each organism that they find in the same area. Lead a discussion on what constitutes a schoolyard community. Identify the term *population* as a way of referring to the different groups of organisms in their study area. For instance, they may find populations of grass, clover, ants, beetles, and worms.

Activities Students identify the different plant and animal populations in their schoolyard community. They discuss what constitutes a community and describe the physical factors involved in the schoolyard community.

TUES

Objective Describing a seashore community.

Teacher Preparation Have available some reference materials showing life along different seashores. If possible, take students on a field trip to the seashore.

Activities Students identify the different plant and animal populations in the seashore community. They discuss some of the physical factors unique to the seashore and adaptations of living organisms to the seashore environment. They apply their knowledge of food chains, food webs, and number pyramids by describing and drawing examples of each. They compare their findings to life and conditions in the previously studied community.

WED

Objective Describing a desert community.

Teacher Preparation Have available some reference materials showing life in deserts. If possible, take students on a field trip into a desert community.

Activities Same type as those on the previous day.

THURS

Objective   Describing a woodland community.

Teacher Preparation   Have available some reference materials showing life in a temperate deciduous forest. If possible, take students on a field trip into a woodland community.

Activities   Same type as those on the previous day.

Notes

FRI

Objective   Describing a pond community.

Teacher Preparation   Have available some reference materials showing life in a freshwater pond. If possible, take students on a field trip to a pond.

Activities   Same type as those on the previous day.

Studying the Human Body Systems: Part I

General Overview Students are naturally curious about their own bodies. They are usually eager to learn more about themselves and how they "tick." The lessons over the next two weeks emphasize the parts and functions of the body systems. These lessons should be taken as only the beginning of many more activities that could be conducted in a study of the human body.

Learning Objectives Students will investigate the skin, skeletal, muscular, digestive, and circulatory systems of the human body.

MON

Objective Investigating the skin.

Teacher Preparation Have available films and reference materials on human skin. Lead a discussion on the function of human skin as a protective covering, as a way of sensing the environment, and as a regulator of body temperature. Discuss the unique characteristics of each person's skin—color, texture, and sensitivity. Have ink pads available for making fingerprints. Provide space on the bulletin board for displaying everyone's fingerprints.

Activities Students participate in identifying the functions of the skin. They make their fingerprints on a white sheet of paper, label it with their name, and place it on the classroom bulletin board. They observe that all fingerprints are different, as are skin colors and textures.

TUES

Objective Investigating bones and the skeletal system.

Teacher Preparation Have available films and reference materials showing pictures of the human skeleton. Obtain a model of the human skeleton to identify the major bones—legs, arms, pelvis, backbone, ribs, breastbone, collar bone, neck bones, and skull. Locate and demonstrate movement in ball-and-socket joints and hinge joints. Point out spaces between backbone vertebrae and fingers and toes that are held together by *ligaments*. Locate some ligaments on chicken bones.

Activities From the model, students locate the various bones in their own bodies.

WED

Objective Investigating muscles and movement.

Teacher Preparation Have available films and reference materials showing pictures of muscles in the human body. Lead a discussion identifying some of the major body muscles. Distinguish between voluntary and involuntary movement of muscles. Describe voluntary muscles as muscles that move when we want them to move, whereas involuntary muscles are not normally controlled by us, as in heartbeat and eyelid movement.

Activities Students feel their voluntary muscles in the arms, legs, stomach, neck, and back when they are tightened. They try to detect which muscles are tight and which are relaxed when they: take a step; do a kneebend; lean forward, backward, and sideways; and move their neck. They run in place for a minute and observe how the diaphragm automatically (involuntarily) increases its movements.

THURS

Objective Describing the functions of the digestive system.

Teacher Preparation Have available films and reference materials showing the different parts and functions of the digestive system. Identify the major organs of the digestive system on a life-size model of the human torso. Have saltine crackers available for each student.

Activities Students locate the major organs in the digestive system and describe their actions on food. They place saltine crackers in their mouths and, without chewing, sense the taste change from salty to sweet and feel the cracker dissolve in their mouths. They apply their knowledge to describe what happens to the cracker after it is swallowed.

FRI

Objective   Investigating the circulatory system.

Teacher Preparation   Have available films and reference materials showing the parts and functions of the circulatory system. Identify the heart, arteries, and veins on a life-size model of the human torso. Describe how blood moves through the heart, throughout the body, back to the heart, and out through the body again. Point out the function of blood in distributing raw materials from digestion and fresh air from the lungs to the body parts, and collecting waste materials from the body and delivering them to the kidneys and lungs. Have them time their pulse rates—a measure of their heartbeats.

Activities   Students participate in the discussion. They time their pulse rates and compare them. They run in place for a minute, time their pulse rates again, and discover that the rate increases as the body takes in more air during the exercise.

Notes

General Overview   As stated in the overview of the previous week, these lessons should be considered only as the beginning for learning about the different body systems. A study of the reproductive system is included within this week. The teacher will, of course, have to determine the most appropriate manner to approach this topic based on knowledge of the students, school, and community. However, we strongly urge that it not be avoided.

Learning Objectives   Students will investigate the respiratory, excretory, nervous, reproductive, and glandular systems of the human body.

MON

Objective   Investigating the respiratory system.

Teacher Preparation   Have available films and reference materials with pictures of the organs in the respiratory system. Lead a discussion identifying the parts and functions of the respiratory system. Refer to a life-size human torso showing the respiratory organs. Have available an empty gallon jar, a deep pan, a ruler taped to the side of the jar, and one drinking straw for every student. Fill the jar full of water and place it upside down in a pan of water.

Activities   Students take turns blowing hard through their straws, and measure how much water they displace with their air. (Refill the jar for each student.) This provides a rough measure of their lung capacity.

TUES

Objective   Describing the excretory system.

Teacher Preparation   Have available films and reference materials with pictures of the excretory organs. Lead a discussion identifying the parts and functions of the excretory system. Refer to a life-size human torso showing the excretory system.

Activities   Students identify the parts of the excretory system and participate in the discussion.

Objective Describing the glandular system.

Teacher Preparation Have available films and reference materials with pictures of the endocrine or glandular system. Lead a discussion on identifying the different glands of the body and some of their functions. Provide each student with a stencil copy of an outline drawing of the human body.

Activities Students locate the glands of the body, mark their positions on the outline of the human body, and label their drawing. They participate in a discussion of each gland's functions.

Notes

WED

Objective Investigating the nervous system.

Teacher Preparation Have available films and reference materials with pictures of the nervous system. Lead a discussion identifying the parts and functions of the nervous system. Refer to a life-size human torso showing the brain, spinal column, and nerve fibers. Describe a simple reflex such as yawning, coughing, and blushing, and a conditioned reflex such as salivating when food is smelled. Have the students demonstrate a simple reflex.

Activities Working in pairs, students demonstrate a simple reflex. One student crosses his legs while the other strikes the leg just below the knee and observes the knee jerk.

THURS

Objective Describing the reproductive system.

Teacher Preparation Have available films and reference materials with pictures of the male and female reproductive systems. Lead a discussion identifying the parts and functions of the reproductive systems in males and females. Invite a medical doctor or family-life specialist to class to lead a discussion on human sexuality geared to the maturity level of the students.

Activities Students identify the parts and describe the functions of the male and female reproductive systems. They participate in the discussion.

Mathematics

Mathematics

1 Recognizing and Using Numerals

2 Learning to Count

3 Studying Number Sequence and Place Value

4 Learning Addition: Two Whole Numbers

5 Practicing Addition: Alternatives

6 Learning Subtraction: Two Whole Numbers

7 Practicing Subtraction: Alternatives

8 Studying Addition and Subtraction

9 Recognizing Geometric Shapes

10 Learning Multiplication

11 Multiplying Two-or-More-Place Numbers

12 Practicing Multiplication: Alternatives

13 Learning Division

14 Dividing Two-or-More-Place Numbers

15 Practicing Division: Alternatives

16 Practicing the Fundamental Processes

17 Telling Time: Minutes, Hours, Days, Weeks, and Months

18 Communicating with Numbers

Recognizing and Using Numerals

General Overview This topic represents almost the beginning in the search for meaning through arithmetic—to learn the numerals that are fundamental to our system of mathematics and to begin to learn to use them. Activities included may be used both for diagnostic purposes and, if diagnosis so indicates, for instruction. Obviously, the teacher must first, at some time, establish the students' level of understanding of arithmetic. If the level is obviously above the tasks, the teacher must continue to diagnose in an effort to arrive at the appropriate level and, in this book, the appropriate topic. For the most part, students must be individually instructed with individual assignments, but there will be occasions when group activity is appropriate and should be used.

Learning Objectives Students will learn to recognize, verbalize, and begin to use numerals.

MON

Objective Learning the names of basic number symbols.

Teacher Preparation Have available a series of cards large enough to print the numerals 1 through 9 and 0 plus the word name of that numeral. For example, on one side of the card print the number "5" and the word "FIVE." On the reverse side draw or glue pictures of that number of objects. If time and equipment permit, tape-record names of symbols. For example: "Now find the card with birds; there are five birds on this card; say the word 'five' and, if you can, write the symbol '5' and the word 'five.'"

Activities Students look at and play with the cards and are asked to learn as many as they can. The teacher goes over each card with the group or with individuals, holding up the card to say the name for the number and to show the reverse side. Later students attempt to draw the symbol.

TUES

Objective Learning to recognize the numbers 1 through 9 and 0.

Teacher Preparation Have available at least ten of one kind of object—for example, ten clothespins, ten cards, ten blocks. Also have available ten cards, each with a number (0 through 9) printed in large and easily read figures.

Activities One at a time, students respond to the following:

1. What number is written on the card that I hold up? (0 through 9)
2. Can you write these numbers on a piece of paper? (Three objects, the flash card 3; four objects, the flash card 4; seven objects, the flash card 7; etc.)
3. Can you write numbers on the chalkboard? (Random)
4. Can you count these objects? (Up to 10) Higher? (Up to 20 or 30)
5. Can you count by memory and without objects? Through 10? Through 20? Higher?

WED

Objective Learning to recognize the numbers 1 through 9 and 0.

Teacher Preparation Have available the following: wood, plastic, cardboard, felt, sandpaper cutouts of the numbers 0 through 9, and a felt board. Also useful are small individual chalkboards.

Activities Using the cutouts for recognition exercises, students are asked to identify the numbers through 9, to feel the numbers, and to trace them with their fingers. They close their eyes and trace the number. Later they write each number they recognize on paper, on their individual chalkboards, or the class chalkboard. They trace each number and then write each number.

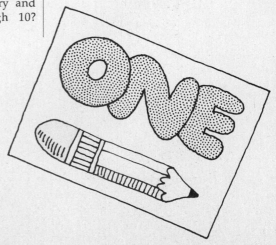

THURS

FRI

Objective   Recognizing and writing the numbers 0 through 9.

Teacher Preparation   Have available a supply of ten or more objects such as dominos, blocks, clothespins, jars, cans, etc. Also have available large printed number cards, one each for the numbers 0 through 9. You may also need your number cutouts (plastic, wood, felt)

Activities   Students participate in the following activities:

1. In two teams similar to a spelling bee, each child has a turn at: (a) verbally recognizing a number symbol—cutout or printed; (b) writing the number on the board; (c) verbally recognizing a number of objects; and (d) writing the number of objects on the board.

2. A game similar to "one-man" circus games: Can you "ring the bell" and answer these? Can you "throw the hoops" and answer? Can you "win a piece of bubble gum" and answer?

3. A one-man game to see who can answer the most arithmetic questions. Johnny got 10, Mary 8, Rosie 9, etc.

Objective   Recognizing, writing, and using the numbers 0 through 9.

Teacher Preparation   Have available a number of trays (food trays, shallow boxes, or something similar) and a variety of objects to place on the trays. You should have up to nine of at least one type of object—for example, nine bottle caps, eight trucks, seven marbles, etc.

Activities   Students place a certain number of objects on trays. The "five" tray (so labeled) gets five trucks this time; the "three" tray, three bottle caps, etc. Write the number on the tray. They are asked to arrange the trays in order—zero through nine. They repeat these activities many times with different objects—five trucks, five bottle caps next, five marbles, etc. Each time they write on the chalkboard or paper the number on their trays.

General Overview Most students come to school with some exposure to numbers: they know some symbols, they can count some, they can verbalize measurement and money, and the like. The teacher needs to find out where students are and select activities appropriate for each. This can be done verbally or in writing by systematically using many of the suggested activities in weeks 1, 2, and 3. Keep in mind that, as with many skills, students need to experience a variety of activities dealing with counting, each as closely related to their background as possible. They need practice, but it must be meaningful practice.

Learning Objectives Students learn to count to 10 (and beyond) and to recognize and write the numerals 0 through 9.

MON

Objective Learning to *want* to count and to begin to count to 10.

Teacher Preparation Have available a variety of articles that students might like to play with—small trucks, boats, or dolls. Use trays to hold the articles. Collect pictures with a variety of objects including, for example, one with two cows, five birds, five members of a family, etc. Have available some large nails, glass eggs, and carton. Make up a series of questions to ask students.

Activities Students respond to the following problems:

1. You are helping your father build a dog house and he asks you to bring him nine nails. Can you do it?

2. You are helping to make a cake and instructions call for four eggs. Can you get four eggs from the carton?

3. Suppose I said that you can play with two trucks. Can you find that number on the tray?

4. You need to find out how much money you have and you have this much (place the same kinds of coins on a tray—nine pennies, eight nickels, or seven dimes, etc.). How many coins do you have?

5. You are told by a friend that you can have five inches of a bubble-gum string. Can you count that many?

Students are shown pictures of a variety of objects and asked to identify the number of objects in each picture.

TUES

Objective Learning to want to count and to count to 10.

Teacher Preparation Have available a variety of sets of things for use with students: eggs, tennis or golf balls, bottle caps, pencils, marbles, blocks, felt cutouts of rabbits or other animals, etc.

Activities Students respond to a variety of situations created by the teacher. How many eggs are in this carton? How many frogs in the pond? Use all numbers, 0 through 10; then try some over 10.

Students are divided into groups to play "baseball." They try to make the highest score by answering correctly the most arithmetic questions, scoring one point for each correct answer. Each correct answer is a "hit" and each incorrect answer an "out"!

WED

Objective Learning to count to 10.

Teacher Preparation Collect a variety of objects (markers) for use in helping students to learn to count—blocks, marbles, bottle caps, beans, ice-cream sticks, stones, toys. Have available the cutouts (wood, plastic, felt), the feltboard, and the individual chalkboards.

Activities Students try to select the correct number in each of the following situations: (1) Select four toys. (2) Give four balls to John. (3) Count out nine bottle caps, seven balls, eight beans, etc. (4) Draw five "moons" on the board. (5) Put seven, eight, or nine felt objects on the feltboard. (6) Count the students in this room. (7) Count the doors. (8) How many chairs - too many to count? (9) How many windows? (10) How many girls? Boys?

THURS

Objective   Learning to count to 10 and above.

Teacher Preparation   In addition to the materials for the previous lessons, have available a number of newspaper advertisement sections, magazine advertisements, and other similar sources of number symbols and cutout possibilities, plus scissors.

Activities   Students participate in the following tasks:

1. Cut out of the newspapers or magazines a certain number of cans of beans or corn—five cans, ten cans, eight cans, etc. They try to find a number in the paper equal to the number of cans—eight cans of corn (paste on a sheet) and the number "8" from an ad. They write these numbers on the board or on paper.

2. Students respond to a variety of comparison situations created by the teacher. How many more in this group (of eight cans) than in this group (of four cans)? How many more red blocks than black? How many more cows in the picture than sheep? How many more footballs than basketballs?

FRI

Objective   Learning to count above 10.

Teacher Preparation   Have available large groups of objects to be used for counting—nails, large beads, bottle caps, popsicle sticks, poker chips, cards, pennies, washers, etc. As always, have available materials on which to write—chalkboards (individual or class), paper, etc. On the trays or table tops arrange groups of objects for students to count—ten on one tray, twenty on another, thirty on a third, etc.

Activities   Students try to count the objects on each tray. If they forget, let them use the method that Alexander used to count his army (see week 3)—that is, for each ten bottle caps, they put a nail (or something) in another pile and then count the nails to give the number of piles of bottle caps. Next they use any number (over 10) of objects on the tray—23, 35, 47, etc. Let them count many, many times. Students write the numbers as they count the various piles and groups. As with all of the activities, they can play games (competition). Use a variety of reward symbols (gold stars, chewing gum, or any other technique that seems to work).

Studying Number Sequence and Place Value

General Overview Early in the study of arithmetic the student has to become familiar with the place-value organization of our number system. This week's activities deal specifically with place value of ones and tens but also brings in activities to help the student learn to understand and use (count) groups of numbers. After students learn to recognize the numbers, write them, and count, they should be about ready to comprehend grouping and place value.

Learning Objectives Students begin to learn about place value of ones and tens and to learn the number sequences of twos, fives, and tens.

MON

Objective Beginning to understand the necessity of place value in a number system.

Teacher Preparation Read about Alexander the Great and how he was supposed to have counted the men in his army. (Briefly, they marched single file past a pot and dropped in a small pebble. When ten pebbles were in the container, a "counter" emptied the pot and dropped a larger stone in another pot. When this pot got ten of the larger stones, another still larger stone was added to a third pot, and so it went—primitive place value!)

Activities Students locate two distinct sizes of pebbles or stones and play "Alexander's Army" (they may need to go through the counter several times).

The teacher varies the technique by creating other situations requiring counting and place value—for example, "You are at the ocean and need to count the fish you catch." Have two kinds of shells for the counters—small ones for the "ones" (through ten) and larger ones (or another kind) for the tens. When students begin to understand, they go to the "hundreds" place with still another shell or stone.

TUES

Objective Learning the difference between "ones" and "tens."

Teacher Preparation Purchase or construct an abacus or counting frame with no more than four "rows," preferably two to start. This can be as simple as a board with two pegs, each of which can hold nine rings exactly. These can be any size, shape, or color.

Activities Students use the abacus to show numbers, 1 through 9 first. Then move to some numbers over 9 trying to show concretely and in writing what happens when you have ten or more beads or rings. In the first place, you do not have enough room on the devices for more than nine but, when you do, it can be shown clearly. (This is a difficult concept and one that is not particularly exciting to students, so the teacher will need to work on it for a short period of time and come back to it often. Use whatever technique for motivation that seems to work.)

WED

Objective Learning the difference between "ones" and "tens."

Teacher Preparation Construct place-value charts for use with individuals and with groups. Have one chart for ones, one for tens, and later one for hundreds. Use slips of paper for ones, groups of ten slips of paper held together with a rubber band for tens, and ten groups of ten packets each for hundreds. With more understanding, you can switch to using different colors for units, tens, and hundreds instead of packets. Show the group how to use the place-value chart.

Activities Students use the place-value chart to place slips in the ones chart—up to the number 9 and then more. The teacher shows them how to gather ten ones in a rubber band and place them in the tens chart. The teacher uses example after example and students practice the technique.

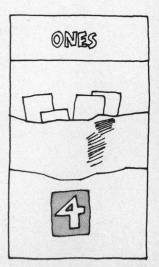

THURS

Objective Learning sequences of numbers.

Teacher Preparation Have available 100 of various kinds of objects, such as nails, washers, bottle caps, cutouts, paper slips, dominos, or poker chips. Place piles of objects on the tray or on a table.

Activities Students try to count the objects on the tray or table. They try to group them in some way to make them easier to count. Eventually they pile them in groups of two and count them. They make up to fifty piles and practice counting. Next they pile in groups of five and count (many times), then ten and count. Finally, they group by threes and count—at least to the number 30.

FRI

Objective Learning sequences of numbers.

Teacher Preparation Have available a set of play money with a supply of nickels and dimes. Locate a variety of things to have in a store with as many pairs or sets for sale as possible.

Activities With a "store" in operation in the classroom, have customers purchase everything either in groups of two or three, paying with nickels and dimes. For example, bubble gum costs 5¢ for two pieces and candy sticks cost 10¢ for three sticks. Students are then forced to use groups, in this case groups of two, three, five, and ten. This can go on for a number of days, perhaps with changing the type of store but still selling by groups.

Students also play games with sequences such as "who can go next—2, 4, 6—now what?" (8, 10, 12, etc.).

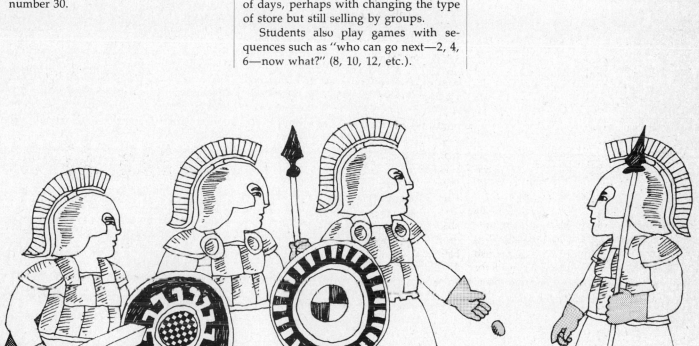

Learning Addition: Two Whole Numbers

General Overview The first of the four fundamental processes to be encountered is addition, and the beginning stage deals with the addition of only two whole numbers or the combinations of any two numbers from 0 through 9. If the student has no knowledge of or concept of addition, all activities in this topic will be in order; however, diagnosis may indicate varying levels of proficiency in addition, in which case some students may need to move ahead to more appropriate activities. There are 100 basic addition facts (sums) to represent the combination of any two numbers from 0 through 9. Eventually students will need to know all of these.

Learning Objectives Students are introduced to the concept of addition and they develop skills to master the addition of one- and two-place whole numbers.

MON

Objective Beginning to see the need for addition and learning to add numbers with sums less than 10.

Teacher Preparation Have available the place-value charts used in earlier lessons, the abacus, a flannel board, and flannel cutouts of a variety of objects—stars, dogs, cats, etc. Also have available your trays and at least ten objects that can be placed on the tray. Make up "stories" about adding numbers (with sums less than 10) so that students can begin to see a need and get a "feel" for adding numbers.

Activities Students respond to questions such as:

1. You are in a candy store and can buy three gum balls and two chocolate drops. How many would you have altogether? Show me on the feltboard; on the trays.

2. I have four pebbles in one hand and three in the other. How many will I have when I put them together?

3. We have four boys and five girls in our group. How many do we have altogether? Show me.

4. Five of us have brown hair and two have black hair. How many people is that altogether? Show me.

5. Can you make up a similar problem?

TUES

Objective Learning to add numbers with sums less than 10.

Teacher Preparation Materials required are the same as for Monday. Emphasis in this lesson is on learning addition facts (specific sums). Use a variety of techniques and approaches to introduce students to the forty-five addition facts with sums less than 10. Use the abacus, place-value charts, flannel boards, chalkboard, and "counting trays." You show them some and have them show you. Always point out certain basic characteristics such as $3 + 2 = 2 + 3$, $4 + 3 = 3 + 4$, etc.

Activities Students play games when appropriate: "Who can show me on the abacus what $3 + 5$ is?—and on the blackboard?—and on the place-value charts?" They have a "math contest" to see who can make the most number sentences ($3 + 2 = 5$) or who can add the most in a row without missing. They use the objects and trays consistently.

WED

Objective Developing facility in the use of the basic addition facts.

Teacher Preparation Use the materials suggested in the previous lesson and add, whenever possible, additional interesting items for counting. Possibilities include seashells; interesting rocks; crayons; beans (a variety); beads; candy, gum, or cookies (can be eaten if kept clean); leaves; nuts or fruit; dominos; dice; bottle tops (variety); small plastic or metal bottles; and cans.

Activities Students participate in situations created by the teacher dealing with sums of more than 10. Using trays or table tops, they count the objects and include, at one time or another, all of the facts. The technique is varied—that is, the teacher shows the students some problems, the students show the teacher some, and students show each other.

THURS

Objective Developing facility in the use of basic addition facts.

Teacher Preparation Use the abacus or counting frame, flannel board, chalkboards, plus objects and trays. One student may pose a problem as shown on the counting frame, another may show it on the flannel board, and another on the chalkboard. In each case, the teacher or students (or both) should verbalize the problem—that is, $3 + 8 = 11$, so three added to eight is eleven and, later, three plus eight equals eleven.

Activities Students work with all of the number pairs at one time or another. In other words, $9 + 1, 9 + 2, . . . ,$ and $9 + 9$ covers all of the "nine" series. Then, $8 + 1$, $8 + 2, 8 + 3$, etc.

FRI

Objective Developing facility in the addition of any two whole numbers.

Teacher Preparation This lesson is to "summarize" addition of all the possibilities of two whole numbers. Have available all the materials used earlier. Make flash cards of all 100 addition facts, with the problem on one side and the answer on the other.

Activities Students use the flash cards in a variety of ways—individually or in groups. They use them in student-student, student-teacher, and teacher-student situations. In all situations, however, the teacher must note mistakes and take the time to locate the source of trouble for each student.

Practicing Addition: Alternatives

General Overview As with most skill areas, practice may be necessary for mastery. With arithmetic the necessity for practice has been realized for years and "drill" has been the technique. Although drill and other mastery techniques will prove effective, care must be exercised in the selection of the technique and the amount of practice required. And as stressed throughout this book, individual diagnosis and prescription are essential, for practice, when not needed, can be devastating and a definite negative learning factor.

This series of suggested activities, as with the other lessons following each of the major skills areas, provides the framework for practice—hopefully meaningful and exciting practice that will help each youngster learn to handle better the skill being stressed.

Learning Objectives Students will attempt to improve the skills involved in addition.

MON

Objective Practicing addition through the use of "math bank" practice cards and ledgers.

Teacher Preparation Create a "math bank" system where students may go to the classroom bank to "borrow" and "repay" when complete. Devise a point system for each "loan" and "repayment" and establish "rewards" for levels of point totals. This department will deal with "addition loans" only and available loans should range from the very simple to the most complex. Begin with "ledgers" (cards or sheets) with several problems involving totals under 10 and only two whole numbers, and develop a sequence of ledgers in addition through the most difficult that you will use. The bank should have "verification" sheets so that students may check their work. Each ledger should have a point value in one corner so that students may accumulate a total. The bank should also maintain a "personalized" account ledger for each student to keep records on accounts paid. Ultimately, the bank should contain several hundred loan options (cards) so that students have many opportunities to practice within the areas of their individual needs.

Activities Students "go to the bank" whenever convenient—individually, in small groups, or in a group as large as can be handled by the teacher and aides. Students will need guidance as to the type of transaction or series of transactions to pursue.

TUES

Objective Practicing addition through the use of "addition football."

Teacher Preparation Refer to week 36 for instruction on playing "arithmetic football." The problem cards to be used in the various decks should be directed to the level of difficulty in addition which students can handle but which will give adequate practice.

Activities Decide on the length of the game before "kickoff"; select teams, elect captains, and "play football!"

WED

Objective Practicing addition through the use of "addition baseball."

Teacher Preparation Refer to week 36 for instructions on playing "arithmetic baseball." The problem cards developed for addition baseball should provide practice opportunities for as many of the players as possible. The four decks should be clearly differentiated in terms of difficulty, with at least one deck within the grasp of the least advanced student.

Activities Decide on the number of innings to be played, create two teams, select a captain, decide on the batting order, and "play ball!"

THURS

Objective Practicing addition through the use of "addition basketball."

Teacher Preparation Refer to week 36 for instructions on playing "arithmetic basketball." Again, the problem cards and various decks developed should allow participation by all players.

Activities Select the teams (any number on a team, but teams should have equal numbers), elect captains, decide on the length of the game and order of play, and begin the game.

FRI

Objective   Practicing addition through the use of "addition road races."

Teacher Preparation   Refer to week 36 for instructions on "arithmetic road racing." When developing the problem cards for the three decks, provide the various levels of difficulty to meet the needs of all youngsters who are playing the game. Spend some time discussing possible race routes, the geography and map location of the course, and other pertinent and interesting information.

Activities   Divide the class into small (three or four) groups; each team should select their race vehicle, draw or locate pictures of their vehicle, and prepare for play. If the entire course cannot be covered, decide on the length of time that the race can last. Then begin the race!

Notes

General Overview   Subtraction is the opposite of addition and is frequently taught along with addition. This can be done with your students if you feel secure in so doing, but generally, to avoid confusion, you may want to deal with each separately first, then later show the relationship. There are 100 basic facts in addition and 100 basic facts in subtraction. Students need to become proficient in using *all* of these facts—orally, in writing, and in relevant situations.

Learning Objectives   Students understand the concept of subtraction and develop skills to master the subtraction of one- and two-place whole numbers.

MON

Objective   Beginning to see the need for subtraction and subtracting numbers under 10.

Teacher Preparation   You need all of the items suggested for addition of two whole numbers plus a set of flash cards with the subtraction facts printed on one side and answers on the other (5 − 3 on one side and 2 on the other). You should have a total of 100 cards. Make up stories about subtracting numbers so that, as with addition, students will begin to see a need for subtraction and also learn to subtract.

Activities   Students respond to questions such as:

1. You have five pennies and you lost two. How many do you have left? Let's find out: here are five pennies; take away two.

2. You lose one tire off your car. How many are left?

3. You have seven library books and take back five. How many do you have left?

4. Six of your socks have holes in them, but you have ten socks in all. Let's see, here are ten socks. Take away six and how many are left?

5. Can you think of other problems where you must take away or subtract?

TUES

Objective   Learning the simple subtraction facts—combinations under 9.

Teacher Preparation   Have available all of the materials mentioned for addition; continually acquire objects as you are able to do so to add variety to "things" available to students. Playing cards should be added to the collection; have students make cutouts out of heavy paper or cardboard. Nearly any object can be used for these exercises. Create situations in which students must "take away" from all numbers, 9 and under; this includes fifty-five facts.

Activities   Students use the trays and all objects in various combinations. Nine cards—"discard" five; four dominos—take away one, then one more, then two more; eight bottle caps—take away five. Students use the flannel board for the same activity. They use the abacus to create problems: nine beads—take away four, etc.

WED

Objective Learning all of the basic subtraction facts.

Teacher Preparation Have available all of the equipment mentioned above and be prepared to create a variety of mathematics situations involving subtraction of the numbers 18 and less and zero. With numbers over 12, you and the students may get tired of using "things," so it may be more convenient to use the counting frame.

Activities Students answer a succession of simple subtraction problems that help to emphasize the relationship of subtraction to addition. For example, $9 + 9 = 18$, so $18 - 9 = 9$; $9 + 8 = 17$, so $17 - 9 = 8$ and $17 - 8 = 9$. They use the counting frame or abacus to answer the problem; they use objects where appropriate. Sixteen marbles less seven marbles leaves nine marbles. They use the trays whenever possible just to keep all of the objects together.

THURS

Objective Developing facility in the use of the basic subtraction facts.

Teacher Preparation Have available the subtraction flash cards and the counting frame.

Activities Students answer a succession of subtraction problems as rapidly as possible. They first use the counting frame and, if they handle the facts by manipulation, move to more abstract techniques. They manipulate the beads on the frame: 8 beads—move 3—how many are left? Then they do the same problems without actually moving or handling the beads. Once they can do this, move to the flash cards and go through as rapidly as possible for as long as the interest remains.

FRI

Objective Developing facility in the use of the basic subtraction facts.

Teacher Preparation This lesson is a "summary session" for all the work in subtraction to this point, and these activities can be used with groups of students or with individuals. Have available the materials used throughout this topic plus a set of flash cards with all 100 basic subtraction facts.

Activities Students use the flash cards in any number of ways—individually, small groups, or large groups. They can be used in student-student, student-teacher, or teacher-student encounters, but in all situations the purpose is to develop facility in the use of basic subtraction facts *and* to identify problems in use as they occur so that the source of the trouble may be remedied.

Notes

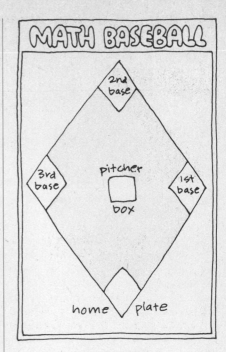

Practicing Subtraction: Alternatives

General Overview   As discussed in the General Overview for week 5, practice is a necessity for most students. The more relevant and interesting the practice is, the more likely students will benefit to the maximum. The games included for subtraction can provide interesting practice alternatives. Often the key to the success of the game is the teacher: prior preparation, understanding of game rules, enthusiasm, and a desire for some fun are all recommended ingredients.

Learning Objectives   Students will attempt to improve the skills involved in subtraction.

MON

Objective   Practicing subtraction through "math bank" practice cards and ledgers.

Teacher Preparation   Refer to week 5 for detailed instructions on using the "math bank." Once the bank is established, it can be used for any of the skill areas—addition, subtraction, multiplication, or division. The problem "ledgers" (cards or sheets) are similar to contracts and a large number are necessary, particularly for groups with wide variation in skill development.

Activities   Students go to the bank whenever possible, at first under rather close guidance of the "head loan office," and later, as they learn the system, without immediate supervision.

TUES

Objective   Practicing subtraction through the "great sailboat race."

Teacher Preparation   Refer to week 36 for detailed instruction on "arithmetic boat races." Use the race course to advantage in the study of geography, measurement, estimating, map reading, and many other areas. Be sure to provide differences in difficulty in the three decks of problem cards.

Activities   Select (appoint or elect) boat crews of two to four members, select a captain for each crew, decide upon the order, and begin the race!

WED

Objective   Practicing subtraction through "subtraction football."

Teacher Preparation   Refer to week 36 and to week 5 for instructions and ideas on playing "arithmetic football." Prepare the decks of cards for subtraction.

Activities   Establish time limits, follow instructions, and play ball!

THURS

Objective   Practicing subtraction through playing "subtraction baseball."

Teacher Preparation   Refer to week 36 and to week 5 for instructions and ideas on "arithmetic baseball." Prepare the decks for subtraction.

Activities   Play baseball!

FRI

Objective   Practicing subtraction through playing "subtraction basketball."

Teacher Preparation   Refer to week 36 and to week 5 for instructions and ideas on arithmetic basketball. Prepare the decks for subtraction.

Activities   Play basketball!

Notes

Studying Addition and Subtraction

General Overview   This unit is intended to extend the student's ability to add and subtract to the extent necessary to funtion in society. They first deal with numbers not requiring regrouping, then move to regrouping in both addition and subtraction. The primary instructional tools for this unit are the place-value chart and the abacus.

Learning Objectives   Students extend their concept of the facility in the use of addition and subtraction with and without regrouping.

MON

Objective   Developing facility in the addition of groups of numbers.

Teacher Preparation   Have available the place-value charts, counting frame, and a variety of objects such as bottle tops, playing cards, and dominos. Make a set of flash cards with combinations that when added do not require "carrying" or regrouping; for example, 14 + 12, 11 + 25, 9 + 30, 8 + 21, 36 + 23, 42 + 7, 53 + 14, etc.

Activities   Students work a variety of addition situations, first with two numbers and then with more, but none requiring regrouping. They use both horizontal and vertical "sentences":

$$12 + 13 = \qquad \begin{array}{r} 12 \\ +13 \end{array}$$

Some students may need to use concrete objects even in this phase, but as soon as possible they should use the number frame and place-value charts and then the flash cards and the chalkboard. The teacher helps students to think in terms of units, tens, and hundreds and to analyze numbers in this regard. Twelve (12) is one 10 and two units; thirteen (13) is one 10 and three units. When the two are added you have two 10's and five units: 13 + 12 = 25. When students can handle this, they can easily use both the place-value chart and the counting frame.

TUES

Objective   Developing facility in the subtraction of numbers.

Teacher Preparation   A variety of objects, the counting frames, and place-value charts are needed in this lesson plus a set of subtraction flash cards. Use subtraction facts of 100 and below unless your students have demonstrated considerable proficiency in handling subtraction. Review first by creating situations (using the counting frame, various objects, chalkboard, flash cards, place-value chart) in which students subtract the number 1 from all numbers through 19, then 2, then 3, etc.

Activities   Following the review session, students subtract two-digit numbers not requiring regrouping:

53 − 20 (5 tens and 3 units take
away 2 tens and 0 units)

$$\begin{array}{r} 53 \\ -20 \end{array}$$ (0 from 3 = 3 and 2 from 5 = 3, or answer of 33)

When possible, they use either the counting frame or abacus to set up the problems, being sure to represent numbers involved as tens and ones or units: 75 is 7 tens and 5 ones; 49 is 4 tens and 9 units; take away 22 (2 tens and 2 ones).

WED

Objective   Beginning to regroup in addition.

Teacher Preparation   Materials needed include only the place-value charts and the counting frame. Have in mind a variety of examples that require regrouping in addition, such as 23 + 28 = 51, 38 + 27 = 65, and 19 + 19 = 38.

Activities   Students work a variety of problems in which regrouping is necessary, such as adding 19 and 18: 1 ten and 9 ones plus 1 ten and 8 ones equals 2 tens and 17 ones (but 17 ones is 1 ten and 7 ones) which equals 3 tens and 7 ones or 37. They illustrate and work problems presented by the teacher using either the counting frame or place-value charts. Students practice on a succession of problems.

THURS

Objective   Developing facility in regrouping in subtraction.

Teacher Preparation   Have available the counting frame, place-value charts, and a set of flash cards with examples utilizing numbers 99 and less and which require regrouping to complete (43 − 28). As with addition, present a variety of examples in which students must regroup. Use the place-value charts to illustrate. Forty-three (4 tens and 3 ones) take away twenty-eight (2 tens and 8 ones). Obviously, some units must come from the only other source—the tens.

Activities   Students work a wide variety of regrouping problems in subtraction, using the counting frame, place-value charts, and chalkboard. When possible, students work problems "in their heads."

FRI

Objective   Developing facility in addition and subtraction.

Teacher Preparation   Have available the devices mentioned earlier in this topic—counting frame, place-value charts, chalkboard, and flash cards.

Activities   Students participate in the following games:

1. Let's go fishing! For every "easy" addition or subtraction, you catch a small fish; for hard ones you catch a larger one. Have a "stringer" with different-size fish to attach. The ones with the biggest fish and largest string are the best at fishing.

2. A trip to the ice-cream store! For each correct answer (addition or subtraction), you get to take three steps toward the store. For really hard questions take five steps. It's 75 steps to the store—who can get there first? Take those steps and keep count!

3. Paper money and how you earn it! For each correct answer you get $1 in play money. Anyone who earns $10 gets to have a five-minute break today!

Notes

General Overview   For thousands of years people have used geometry to solve problems, to design construction projects, and to please the eye. Geometric shapes and designs abound, and we use them in all kinds of activities and transactions in our contemporary world. This week's activities focus on geometric shapes and some of their characteristics.

Learning Objectives   Students learn to work with geometric shapes.

MON

Objective   Learning about the square.

Teacher Preparation   Find a number of rulers, English or metric, and protractors or other instruments to measure a 90-degree angle. Have available construction paper, pens, and scissors. Describe a square for the students and try to locate squares in the classroom, school, and community. Talk about how to make a square.

Activities   Using the ruler and protractor, students mark off squares of various sizes. They measure the size and try to compute the area. They help make a bulletin board of squares (various colors), showing the length of each side and the area. Some students may want to make a three-dimensional square or cube. Others may want to make squares from toothpicks or other similar articles.

TUES

Objective   Learning about rectangles.

Teacher Preparation   Have available the same materials as in the previous lesson plus a tape measure of 50 feet or 100 feet. Identify and discuss the shape of a rectangle.

Activities   Students identify some rectangles within the school grounds and measure them. They begin with things within the classroom—books, notebook paper, notebooks, the teacher's desk, windowpanes, windows, doors, the classroom itself, and the like.

Students use construction paper of various colors to make rectangles. They help prepare a bulletin board of rectangles and they make rectangles with toothpicks, sticks, popsicle sticks, and other suitable materials.

They answer: "What is the distance around your rectangle?"

WED

Objective   Learning about circles.

Teacher Preparation Have available several compasses or round objects (plates, etc.) that can serve as patterns for circles. Have several different colors of construction paper, scissors, glue, and felt-point pens. Talk with students about circles and what makes them unique from any other geometric shape. Think of examples of circles that people use and find as many as possible. Examples are glasses, dishes, wheels (in motors, machines, watches), hats, buildings, and food (pastries, vegetables and fruits, etc.)

Activities Students construct circles of different sizes on various colors of paper and post them on the bulletin board. They answer these types of questions: How big are they? Diameter? Radius? How far is it all the way around a circle?

THURS

Objective   Becoming familiar with triangles.

Teacher Preparation Have available the equipment and materials suggested for Monday and Tuesday with different colors of poster paper, scissors, glue, toothpicks, tongue depressors, lollipop sticks, straws, and similar materials to construct triangles. As with the other shapes, discuss the triangle and what makes a figure a triangle.

Activities   Students try to locate some triangles in the classroom, around the school, and other places they may recall ("yield" highway signs, auto windows, arrowheads). Students construct triangles—some with all sides the same length and some with only two sides the same—out of construction paper or glue together toothpicks, soda straws, or other suitable sticks.

FRI

Objective Constructing and using geometric shapes.

Teacher Preparation Have available all of the equipment mentioned in Monday through Thursday's lessons plus ample construction materials.

Activities   Students construct the four geometric shapes. They make posters, bulletin boards, mobiles, or buildings using soda straws, sticks, paper, etc.

Notes

Learning Multiplication

General Overview Multiplication is one of the four fundamental arithmetic processes. All four of the processes are important in contemporary "survival," and students should be able to multiply with sufficient skill to handle purchasing and other personal and vocational activity.

Learning Objectives Students learn to multiply.

MON

Objective Beginning to learn the concept and techniques of multiplication.

Teacher Preparation Have available a variety of items such as checkers and bottle caps for use in grouping, perhaps up to fifteen for each student or pair. Have available strips of paper with geometric designs in various numbers. For example, have some strips with three circles, some with three squares, four circles, four squares, etc.

Activities Students become involved in problem situations. Place six blocks on the table and ask how many are there? Arrange the six into two separate groups with three in each—now how many are there altogether? Two groups of three each equals two threes. So two times three is or equals six. On the board:

$$3 \times 2 = 6 \quad \text{and} \quad \begin{array}{r} 3 \\ \times 2 \\ \hline 6 \end{array}$$

Now rearrange into three piles of two blocks each:

$$2 \times 3 = 6 \quad \text{and} \quad \begin{array}{r} 2 \\ \times 3 \\ \hline 6 \end{array}$$

Do this a number of times with each student, using 2×2 through 2×9. Students experiment and practice with real objects.

TUES

Objective Learning to multiply with 2, 3, 4, and 5.

Teacher Preparation Make sets of strips of paper long enough to accommodate sets of geometric designs of varying numbers. For example, make strips one inch by three inches with three circles on each; make some of same size with three squares. Make some one inch by four inches with four circles, then with four squares, and so on through the number 5.

Activities Students use the strips to continue with multiplication. They repeat many of the situations used in the preceding lesson with items, this time using strips (markers): two markers each with three designs, two markers each with seven designs, etc. They try the markers vertically and horizontally and write, for example:

$$2 \times 3 = 6 \quad \text{and} \quad \begin{array}{r} 2 \\ \times 3 \\ \hline 6 \end{array}$$

WED

Objective Learning to multiply with 6, 7, 8, and 9.

Teacher Preparation Make more strips with shapes to include 6, 7, 8, and 9. Have enough to show a number times itself (8×8, 9×9).

Activities Using the strips as in Tuesday's lesson, students solve multiplication problems utilizing 6, 7, 8, and 9. In all cases the teacher writes the problems on the board in different forms:

$$8 \times 8 = 64 \quad \text{and} \quad \begin{array}{r} 8 \\ \times 8 \\ \hline 64 \end{array}$$

THURS

Objective   Learning to utilize the 90 basic multiplication facts.

Teacher Preparation   Make a set of flash cards of the 90 basic multiplication facts. Have the problem on one side (8 × 8) and the answer on the other (64).

Activities   Students use the flash cards to practice with the multiplication facts, either individually or in groups. Groups play team games such as math baseball, math basketball, math football, and the like (see week 36). The teacher could set certain goals (10 correct with no "misses") with specified rewards.

FRI

Objective   Learning the basic multiplication facts.

Teacher Preparation   Prepare some cards that present some real situations in which students *need* to multiply (rather than add). For example: If you buy 9 pieces of gum at 2¢ each, how much money will it cost? Eight candy bars at 9¢ each? Four books at $5 each? Nine knives at $3 each?

Activities   Students use the cards or charts with real situations along with the flash cards prepared for Thursday to provide more practice opportunities. Again, students play games or participate in "challenge" situations if this appears appropriate.

MATH FOOTBALL

Multiplying Two-or-More-Place Numbers

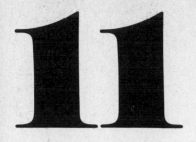

General Overview This topic continues to provide opportunities for students to learn to multiply with at least two- and three-place numbers. As with other basic operations, meaningful practice is a must.

Learning Objectives Students learn to multiply with two-or-more-place numbers.

MON

Objective Learning to multiply with a one-place number.

Teacher Preparation Required will be materials used earlier in developing the idea of place value: place-value charts, counting frames, or abacus. Begin this lesson with numbers like 14 times 2 (1 ten and 4 ones times 2:

$$\begin{array}{ccc} 10 & 4 & 14 \\ \underline{\times 2} + \underline{\times 2} = 28 & \text{or} & \underline{\times 2} \\ 20 & 8 & 28 \end{array}$$

or, if necessary, show two groups of 14 and let them count. Next, 14×3, requiring carrying. With the place-value chart, show 14 as "1 ten and 4 ones" times 3, so make three sets of "1 ten" and "4 ones." Then count "3 tens" and "12 ones." But "12 ones" is "1 ten" and "2 ones," so the product of 14×3 is 42 or "4 tens and 2 ones."

Activities The teacher and students set up a variety of situations utilizing two-place numbers multiplied by a one-place number (up to 99). In addition to using objects, place-value charts, and the abacus, problems are written on the board and students write them both on the board and on paper.

TUES

Objective Learning to multiply using one-place numbers.

Teacher Preparation The same type of materials are required for this lesson as for the one preceding; examples should be with three-place numbers. Begin with numbers like 222×2 (no carrying) and establish understanding of the process. Two hundred and twenty-two is $200 + 20 + 2$. This multiplied by 2 is $400 + 40 + 4$ or 444; $222 \times 2 = 444$. Another example: $597 \times 3 = (500 \times 3) + (90 \times 3) + (7 \times 3) = 1500 + 270 + 21 = 1791$.

Activities Students practice the problems and begin to increase the numbers until they work through 9.

WED

Objective Learning to multiply with one-place numbers.

Teacher Preparation Make a set of flash cards to give practice with two- and three-place numbers multiplied by a one-place number. Print the problem on one side and the answer on the other.

Activities Students use the flash cards to practice the multiplication problems. They play games, set up expectations and rewards, or use other organizational techniques that work with the group.

THURS

Objective Learning to multiply with two-or-more-place numbers.

Teacher Preparation Have available a number of examples of simple two-place multiplying: 22×12; 25×11; 24×12; 33×13; 34×12; 44×2; 41×13; 31×14; 22×13; etc. Use cards, posters, overlays, or worksheets.

Activities Students are shown what is happening when they multiply 22×12 and get 264: $10 \times (20 + 2) + 2 \times (20 + 2) = 200 + 20 + 40 + 4 = 264$ or

$$\begin{array}{lr} 22 & \\ \underline{\times 12} & \\ 44 & (2 \times 22) \\ \underline{220} & (10 \times 22) \\ 264 & \end{array}$$

Other examples of this type are used, breaking them down as shown and explaining when necessary. All problems are written on the chalkboard and students write and solve them on paper.

FRI

12

Objective Learning to multiply many different kinds of numbers.

Teacher Preparation Make another set of flash cards, this time using numbers that continue the sequence from the set used in Wednesday's lesson. Use two-place times two-place, two-place times three-place, and three-place times three-place numbers.

Activities Students use the flash cards from Wednesday and the ones prepared for this lesson to practice multiplying. Students use the chalkboard, their own paper, or overhead transparency material. They have "challenge" matches, using the cards each time.

Notes

General Overview The alternatives for practice through gaming are unlimited, and those mentioned in this book are only some of the many that will work with youngsters. The rules, instructions, and suggestions are not sacred; they should all be altered if the need arises to make them more useful to students and teachers.

Learning Objectives Students will improve the skills involved in multiplication.

MON

Objective Practicing multiplication skills through "math bank" practice cards and ledgers.

Teacher Preparation Refer to week 5 for instructions on the "math bank" and how to organize it for use in the classroom. Prepare the ledgers for multiplication; as with the other skill areas, a large number will be necessary to provide the variety that must be available for the bank to be effective.

Activities Students use the bank as often as possible; after becoming familiar with the system, they should be able to "bank" on their own.

TUES

Objective Practicing multiplication through "sailboat racing."

Teacher Preparation Refer to week 36 for instructions on planning and organizing "arithmetic sailboat races." Be innovative in setting the course: around the world is but one alternative. Other possibilities: the Great Atlantic Ocean Race; the Gulf Race; the Great Lakes Race; the Seattle to San Diego Race, or the Mississippi River Race. As with all games, be sure to provide a wide selection of problems.

Activities Select crews and a captain, discuss the rules, and get the race started.

WED

Objective Practicing multiplication through "multiplication football."

Teacher Preparation Refer to week 36 for detailed instructions for "arithmetic football." Prepare the decks.

Activities Play football!

THURS | FRI

Objective Practicing multiplication by using "multiplication baseball."

Teacher Preparation Refer to week 36 for instructions on playing "arithmetic baseball." Prepare the necessary decks of problem cards.

Activities Play ball!

Objective Practicing multiplication by using "multiplication basketball."

Teacher Preparation Refer to week 36 for instructions on playing "arithmetic basketball." Prepare the recommended decks of problem cards.

Activities Play ball!

General Overview Division is the last of the fundamental processes of arithmetic to be presented to students, and it will likely be considered the most difficult. Division is frequently taught along with multiplication because of the closeness of the two operations. If you feel your students can handle the two, teach them together; you will probably find it preferable, however, to separate the two while referring back to the multiplication facts to provide a point of reference. Activities during this week introduce the basic division facts.

Learning Objectives Students learn to divide.

MON

Objective Seeing the need for division and beginning to divide.

Teacher Preparation Have available a number of objects that can be "divided," such as playing cards, dominos, sticks, bottle caps, and the like. Make a number of "markers" as used with multiplication: strips of paper with two circles each, three circles, four circles, five circles, and six circles.

Talk to the students about dividing things up—at home, at school, or elsewhere. How would you divide some candy? Some apples, a pie, or money? Suppose you have nine pieces of candy and there are three of you. How would you divide it equally? Let's try (one for you, one for you, and one for you—until each has three). Now you have six pieces; now twelve; now fifteen—try it.

Activities Students use the cards or other objects and divide them evenly among a certain number of others. They do this until they can consistently execute the division of items.

TUES

Objective Beginning to divide.

Teacher Preparation In addition to the markers prepared for Monday, make a number of cards with dots that can be divided. For example, have a card with fifteen dots: and on the reverse have how many fives $(15 \div 5)$, how many threes $(15 \div 3)$, how many ones $(15 \div 1)$; twenty-one dots $(21 \div 3, 21 \div 7)$; twelve dots $(12 \div 2, 12 \div 3, 12 \div 4)$, etc. Have a variety of this type of card available.

Activities Students are presented with simple division problems: $4 \div 2, 6 \div 2, 6 \div 3, 8 \div 2, 9 \div 3, 10 \div 5, 10 \div 2$, etc. They use markers (strips with designs) and the cards with 4, 6, 8, etc., dots and problems.

(For students who have difficulty with these, go back to the "objects" division and let them work both ways, with objects and with symbols. For each problem the teacher *and* student should write out the situation—$6 \div 3$ and $3\overline{)6}$—to begin to get a feel for handling "abstract" division. Division facts may follow the same pattern as multiplication facts, so use a chart to work from simple to complex. In other words, have 1 through 9 across the top of the chart and also down the left side; then, reading across and then down, the intersection gives the fact: $4 \times 9 = 36$, $36 \div 9 = 4$, and $36 \div 4 = 9$.)

WED

Objective Learning the division facts.

Teacher Preparation Make additional division flash cards to be used either with large groups, small groups, or individuals: the problem on one side, the answer on the other. Use cards for students to show the answer (each student would need a set of cards with numerals 0 through 9), let them respond orally, or, if practice in forming numerals is needed, let them write the answer on paper or on the board.

Activities Students continue to practice division, working in large and small groups and individually. Pairs of students are asked to work together—at times two of the faster students and at other times a fast and slow student.

1	2	3	4	5	6	7	8	9
2	4	6	8	10	12	14	16	18
3	6	9	12	15	18	21	24	27
4	8	12	16	20	24	28	32	36
5	10	15	20	25	30	35	40	45
6	12	18	24	30	36	42	48	54
7	14	21	28	35	42	49	56	63
8	16	24	32	40	48	56	64	72
9	18	27	36	45	54	63	72	81

THURS

FRI

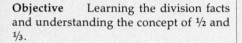

Objective   Learning the division facts and understanding the concept of ½ and ⅓.

Teacher Preparation   Have available the materials listed for Monday, Tuesday, and Wednesday, and make some additional flash cards with ½ of _____ and ⅓ of _____.

Continue with variations of presentation to help students learn the basic division facts. In addition, work on the fact that ½ is the same as dividing by 2 and ⅓ the same as dividing by 3 and vice versa. Present many real problems: each of you gets ⅓ of the chewing gum (6 pieces); ½ of the 4 candy bars; ⅓ of the class gets to stand up. Each time see that they realize that they get the same answer if they divided by 2 or 3.

Activities   Students use the flash cards, do board work, play games, and use other techniques provided by the teacher to gain proficiency in this type of division problem.

Objective   Learning the basic division facts.

Teacher Preparation   Have flash cards for all 90 division facts and make cards or charts showing the ways to write the facts: $3\overline{)15}$, $15 \div 3$, ⅓ of 15, and $^{15}/_{3}$. Label the various parts with the correct terms: divisor, quotient, and dividend.

Activities   Practice sessions are held in groups or individually with all 90 division facts. The teacher writes on the board as many of the problems as possible to give students practice in seeing *and* writing number problems. The "term charts" are used to introduce the division terms and students are encouraged to use the proper terms as they work with division problems.

Dividing Two-or-More-Place Numbers

General Overview   After mastering the basic division facts, students need to learn to handle more complex division problems with more numbers and requiring regrouping. This week's activities introduce students to this type of division and provide practice opportunities.

Learning Objectives   Students learn to divide using two-or-more-place numbers.

MON

Objective   Learning to divide by a one-place number with a two-place dividend.

Teacher Preparation   Have available objects to use in illustrating the problem and also a place-value chart. Make flash cards with a variety of one-place division problems not requiring regrouping ($22 \div 2$).

Activities   The teacher uses one flash card to introduce the "bigger" division problems: $24 \div 2$; $2\overline{)24}$; ½ of 24; ²⁴/₂. Students use playing cards (or any other objects) to solve the problems. The teacher uses the place-value chart to show 24 as 2 tens and 4 ones. The teacher shows on the board: $2\overline{)24} = 2\overline{)20+4} = 2\overline{)20} + 2\overline{)4} = 10 + 2 = 12$ and $2\overline{)36} = 2\overline{)30} + 2\overline{)6} = 15 + 3 = 18$. (Use many examples on the board, then use the flash cards.) Students write and solve the problem on paper, on the board, and then verbally whenever possible.

TUES

Objective   Learning to divide: long and short division.

Teacher Preparation   Have the flash cards available from the preceding lesson. Make a chart showing the steps of long division, in which *all* the numbers used are written down. For example:

$$36 \div 2 = 2\overline{)36} = \frac{36}{2} = \begin{array}{r} 8 \\ 10 \\ \hline \end{array} >18$$
$$\begin{array}{r} 2\overline{)36} \\ 20 \\ \hline 16 \\ 16 \\ \hline \end{array}$$

and

$$24 \div 2 = \begin{array}{r} 2 \\ 10 \\ \hline \end{array} > 12$$
$$\begin{array}{r} 2\overline{)24} \\ 20 \\ \hline 4 \\ 4 \\ \hline \end{array}$$

Activities   Students are given examples of division of medium complexity such as the even-numbered 20's, 30's, 40's, 50's, and 60's. Students should solve these as shown above—by long division—writing down *all* the numbers.

(As students gain proficiency in this method, they may take "shortcuts" and use "short division" where the process is shortened because many of the steps are performed mentally.)

WED

Objective   Learning to divide with a remainder.

Teacher Preparation   Have available flash cards with examples requiring division by a one-place number into numbers under 100 and resulting in a remainder. Examples: $2\overline{)27}$, $3\overline{)34}$, $5\overline{)59}$, $8\overline{)98}$, $7\overline{)85}$.

Activities   The teacher works through several problems on the board, with place-value charts or a counting frame, showing that there *is* a remainder. Example:

$$\begin{array}{r} 1 \\ 10 \\ \hline \end{array} >11 \text{ r } 1$$
$$\begin{array}{r} 3\overline{)34} \\ 30 \\ \hline 4 \\ 3 \\ \hline 1r \end{array}$$

Using the flash-card example, students work several other examples. They work on the board and on paper, using objects, place-value charts, and/or counting frames.

THURS

Objective   Learning to check division problems.

Teacher Preparation   Make a chart showing several examples of checking division problems of the type they have been working: $4\overline{)8} = 2$ and $2 \times 4 = 8$; $2\overline{)24} = 12$ and $2 \times 12 = 24$; and $3\overline{)85} = 28$ r 1 and $3 \times 28 = 84 + 1 = 85$. Explain that to check division you use multiplication because multiplication and division are opposite operations. Have available the division flash cards that you have been using in the previous lessons.

Activities   The teacher shows students how to check division and discusses with them again the relationship of multiplication and division (opposites); use charts to help get it across. Students work other examples using the flash-card examples from preceding lessons. They explain what they did and how they did it.

FRI

Objective   Learning to divide larger numbers.

Teacher Preparation   Have available charts or flash cards of division examples using two-place division, but begin with a divisor that is a multiple of 10 such as 20, 30, 40, and the like. Have two-place and three-place numbers, up to 90, then 110, 120, 130, etc.

Activities   The teacher shows students how to divide $20\overline{)60}$. How many 20's are there in 60? Explain ways to find out: subtract by 20; add by 20; multiply 20 by numbers until you get to 60, etc. Students try other examples: $20\overline{)80}$, $30\overline{)60}$, $40\overline{)80}$, $30\overline{)90}$, etc. Then move to $20\overline{)120}$, $40\overline{)160}$, etc. Then use examples in which the number divided is not a multiple of the divisor: $20\overline{)43}$, $20\overline{)57}$, $30\overline{)85}$, etc., still using the divisor multiple of 10. Next use examples that do not use a multiple of 10: $31\overline{)62}$, $31\overline{)124}$, $31\overline{)95}$, $31\overline{)75}$, etc. The teacher continues to increase the complexity as far as the students can go.

Notes

General Overview   Practice in the skill areas must be a continuing activity, to some extent even after mastery, and it cannot be programmed with assurance that students will reach a certain level at a certain time. For this reason, practice (drill) cannot be scheduled for any one day, week, or month. It must be used as an instructional technique as indicated through assessment of performance. The alternatives suggested in this week and the other weeks for other skill areas could and should occur anytime—again, as needed.

Learning Objectives   Students will improve the skills involved in division.

MON

Objective   Practicing division through "math bank" practice cards and ledgers.

Teacher Preparation   Refer to week 5 for instructions on the organization and operation of the "math bank." Prepare the ledgers for division, including a large number to provide the variety and range necessary for most groups.

Activities   Use the bank for individuals or groups. If the bank has been used for the other skill areas, students should now be able to function easily alone in the bank.

TUES

Objective   Practicing division through "division football."

Teacher Preparation   Refer to week 36 for instructions on "arithmetic football." Prepare the decks as instructed, being sure to provide an adequate range and variety for students in the group.

Activities   Play ball!

Practicing the Fundamental Processes

16

WED

Objective Practicing division through the use of "division baseball."

Teacher Preparation Refer to week 36 for details on game organization and operation. Prepare the decks of problem cards.

Activities Play ball!

THURS

Objective Practicing division through the use of "division basketball."

Teacher Preparation Refer to week 36 for instructions on how to plan and organize for "arithmetic basketball." Prepare the decks of problem cards.

Activities Play ball!

FRI

Objective Practicing division through the use of "division sailboat races."

Teacher Preparation Refer to week 36 for instructions on planning and organizing for sailboat races. Prepare the decks of problem cards.

Activities Get the race underway!

Notes

General Overview As has been consistently emphasized in this book, practice for skill development in the fundamental processes of arithmetic is essential, but it must be challenging and interesting so that a positive attitude toward arithmetic is encouraged. The games presented in week 36 and used after each of the fundamental processes are suggested because they can be fun, challenging, and useful in learning. They are utilized here for more opportunities for meaningful practice in solving problems involved with the basic processes of arithmetic.

Learning Objectives Students practice all four of the fundamental processes through playing arithmetic games.

MON

Objective Practicing solving problems in addition, subtraction, multiplication, and division through playing "arithmetic football."

Teacher Preparation Refer to week 36 for ideas and instructions. If "arithmetic football" has been used for the various skill areas (addition, subtraction, multiplication, and division), the decks of problem cards from each of those can be combined and shuffled to form the decks for this game. If not, new decks with problems from the four areas must be developed.

Activities Play ball!

TUES

Objective Practicing solving arithmetic problems through playing "arithmetic baseball."

Teacher Preparation Refer to week 36 for specific instructions on "arithmetic baseball." Use the decks of problems from the other weeks for this game—mix them up (shuffle) before play begins.

Activities Play ball!

WED

Objective Practicing solving arithmetic problems through playing "arithmetic basketball."

Teacher Preparation Refer to week 36 for specific instructions. Use the decks of cards from the other weeks for this game; be sure to shuffle thoroughly.

Activities Play ball!

THURS

Objective Practicing solving arithmetic problems through "vehicle racing."

Teacher Preparation Refer to week 36 for specific instructions on "arithmetic racing." Use decks of cards from preceding "racing" games and prepare other cards to include all areas of interest (addition, subtraction, multiplication, and division).

Activities Begin the race!

FRI

Objective Practicing solving arithmetic problems through "sailboat racing."

Teacher Preparation Refer to week 36 for instructions on "sailboat racing." Prepare decks of problem cards to cover all skill areas of concern.

Activities Begin the race!

Notes

17

General Overview Another essential involving mathematics is telling time. Without facility in this area, one will have a multitude of difficulties simply in surviving in our contemporary society. As in most mathematics areas, you will need to find out how much each student knows about telling time and then begin at an appropriate point.

Learning Objectives Students learn to tell time using minutes, hours, days, weeks, and months.

MON

Objective Learning about the clock and how it works to keep time.

Teacher Preparation Have available as many different types of clocks as possible —large, small, simple, and complex. Try to have at least one digital clock, too. Make or purchase large clock faces (perhaps 10–12 inches in diameter) with hands that will move. Try to locate enough to let each youngster have one to use. Make a large demonstration clock for you to use.

Activities Students participate in a discussion of the various clocks that are on display. They talk about the shapes, where they came from, the alarms (if they have alarms), and try to see what makes them work. They talk about clocks they have at home; they look at a digital clock, if available; and they discuss sundials. They bring in pictures of clocks for a colláge or bulletin board.

Students are given cardboard clocks and shown how the hours of the day look on a clock face. At each hour they duplicate the time on their clocks. Next they are shown thirty-minute intervals—30 minutes past 3 or 3:30 (A.M. and P.M. will be explained later, but use it each time you have a chance). Next they are shown fifteen-minute intervals: 8:15, 12:15, etc.

TUES

Objective Learning to tell time.

Teacher Preparation On the large demonstration clock mark the minutes, 1–60, clearly. Make a set of flash cards, each with a clock face and the hands drawn to some particular time. Have some that show the hours (1–12), some with the half hours, some with quarter and three-quarter hours, and a variety of other times. Have the clock face on one side and the time written in large numbers on the reverse.

Activities Students observe the minutes marked on the face of the big clock. They see the minute markings on most other clocks and count them. They notice that most clocks mark in five-minute intervals, although some use fifteen-minute intervals.

Using the clock faces on the flash cards, students create the time shown on their own clocks. After facility with this, they read the time from the reverse side of the card and, again, show it on their clocks. They start with the hours, then half, quarter and three-quarter, and then various other times.

12:00 A.M. 1:00 A.M. 2:00 A.M. 3:00 A.M. 4:00 A.M. 5:00 A.M. 6:00 A.M. 7:00 A.M. 8:00 A.M. 9:00 A.M.
MIDNIGHT

1:00 P.M. 2:00 P.M. 3:00 P.M. 4:00 P.M. 5:00 P.M. 6:00 P.M. 7:00 P.M. 8:00 P.M. 9:00 P.M.

WED

Objective   Learning about A.M., P.M., and hours in a day.

Teacher Preparation   Make a large "time line" showing the hours in a day. Begin at 12 midnight and continue through 12 midnight twenty-four hours later, marking each hour with the number corresponding to the time of day. Collect a variety of pictures from magazines and newspapers with events that can easily be categorized as happening in the morning, the evening, or at night. Examples: the sun rising, eating breakfast, dressing in pajamas, going to bed, etc.

Activities   Students examine the pictures that have been collected and tell what time they think it is. When they say "9 o'clock," they are asked if it is morning or afternoon, A.M. or P.M. They are told the meaning of A.M. (ante meridiem) and P.M. (post meridiem) and discuss when each occurs.

On the time line, the students help to "fill in"—create a dark sky for the night hours, perhaps with a moon; sunrise, morning sun; afternoon sky, sunset; and dark again—including various school activities such as when school begins, lunch, recess, end of school day, etc. They mark A.M. and P.M. (With some students you may want to tell time as used by some pilots, the military, and some other agencies: 1 through 24.)

THURS

Objective   Learning the days of the week.

Teacher Preparation   Make cards with the days of the week printed in large letters, including Saturday and Sunday. Make a second set with string attached so they can be hung around the neck of a student. Make a set of flash cards with the names of the days printed in large letters. Have available magazines and newspapers that can be torn apart.

Activities   At designated locations in the room, students post each day the name of the day of the week. Students should be assigned this task; on Friday post also Saturday and Sunday.

Students participate in a story-time game. Seven students hang cards around their necks with one of the days of the week. In the story, the various days of the week are mentioned. The student must hop up and say the name on his or her card when it appears in the story. (The story must include all days and can go on and on. Let youngsters take turns or have enough "day cards" for each to have one —you may have three Mondays, three Tuesdays, etc.)

Play games with the "day" flash cards, with students naming them as the teacher holds up cards and asks questions such as: "Which day comes before?" "Which day comes after?"

Students keep a day-by-day diary chart, with the day, the weather, and important events.

FRI

Objective   Learning the days of the month.

Teacher Preparation   Collect a variety of calendars, preferably large, colorful ones. Make a stencil so that you can run off a number of blank calendars. Make a series of cards with "1" through "31" to represent the date (these are to be used with the days of the week: Monday, October 9) plus cards with the twelve months.

Activities   Talk with students about the seven days in a week and then about the number of days in months: some 30, some 31, and February. Show them the calendars and let them count the days in each of the twelve months. Find the present month and talk about the number of days in that month. Which week? How many days have gone by? How many left?

Students use the stencil and make a calendar for that month. (The teacher may want to give each youngster several calendars: one for recording the weather, one for important events, one for birthdays, and the like.)

Each day one student posts the date on the board near the day of the week and the month.

10:00 A.M. 11:00 A.M. 12:00 P.M.
NOON

10:00 P.M. 11:00 P.M. 12:00 A.M.
MIDNIGHT

Notes

18

General Overview Through numbers we communicate and, in order to communicate most effectively, we must all be able to give and interpret messages through numbers. There are zip codes, house numbers, road signs, license tags, phone numbers, social-security numbers, account numbers, model and serial numbers, numbers in consumer information, and many more. This week focuses on number messages essential for successful living.

Learning Objectives Students recognize the communication potential of numbers and learn to use them.

MON

Objective Learning to use street (house) numbers, box numbers, and/or route numbers

Teacher Preparation Locate maps of the town(s) in which most students in the class live or, if a rural area, the closest town plus a map showing roads in the rural area. Discuss with students how the buildings are numbered—whether house numbers or box and route numbers—and point out patterns such as odd numbers on one side and even on the other and progressing from small to large in certain directions.

Activities Students bring their addresses and the addresses of several people they know. On the maps of the region they locate the home of each student and draw it in, including the house number. They write in the friends' addresses. They draw local businesses, farms, and significant buildings (museums, schools, auditorium, city hall), including the number addresses. They discuss any noticeable patterns. (Students may want to make their own model town for a variety of areas of study and, if this is feasible, mark block, streets, house numbers, and the like.) The teacher gives students certain addresses to locate on the maps. (If possible, take the students on a field trip around the town or region to show them the numbering system.)

TUES

Objective Learning to use number signs on highways.

Teacher Preparation Bring in as many magazines and newspapers as possible for use in locating road signs using numbers (speed limits, distances, and route numbers). Some very good literature can be obtained from state transportation and licensing agencies. Have available poster paper; yellow, red, and black construction paper; glue; and felt-tip pens. Make a set of flash cards of road signs, with each card showing a number message on a sign of the shape and color used in the vicinity.

Activities Students look through the magazines and other material collected and cut out pictures of road-sign messages. They display these on the bulletin boards or make collage-type posters. They also make model road signs with number messages, such as speed limits (55 mph maximum speed; school—15 mph), distance (North Pole—15 miles), and route numbers (U.S. 66). Students use the flash cards individually or in groups.

FRI

Objective Learning to use phone numbers, catalog numbers, and model or serial numbers.

Teacher Preparation Make charts showing the local access-code numbers (if used in the area), the local area code, and the sequence of events when making local or long-distance calls. Try also to locate an old phone to use in the classroom. Obtain catalogs and order forms from a local firm. Xerox the order forms in quantities sufficient for each student to use several.

Activities The teacher discusses the access-code numbers, area-code numbers, and phone numbers as shown on the charts. Each student is given a number and a chance to dial on a real or play phone.

Each student selects items from the catalogs available and fills out an order form, using numbers for the address, zip code, and phone, as well as the catalog numbers.

Each student looks up, at home, the serial or model number of at least one appliance and brings it in to record on a master list at school. They discuss the use and value of serial numbers (for warranties, parts replacement, identification in case of theft, etc.).

Notes

WED

Objective Learning about numbers connected with automobiles.

Teacher Preparation Three major types of activity are suggested in this lesson, with two requiring permission from car owners. Obtain permission (and keys, if necessary) from about five car owners whose cars will be in the school parking lot. Have available poster paper, pens, and lined sheets of notebook paper for youngsters to use in recording data.

Activities Students visit the parking lot and write down the license numbers of a variety of cars, keeping a list such as: blue Ford, license #____. Then they get inside the car and look at the mileage indicator, writing this down also. Next, they open the hood and locate the serial number of the car or motor and write this down. Then they discuss the significance of all of the numbers: What does each do? Who uses them—the owner, the police, the license bureau, car dealers?

THURS

Objective Learning to use zip codes.

Teacher Preparation Obtain envelopes that have been mailed to people in your community and also from neighboring communities. Obtain from the post office (or school office, perhaps) a directory of zip codes. Locate enough envelopes so that each student can have one and cut out plain white paper to resemble the front of an envelope.

Activities Students bring a used envelope from home and discuss the addresses, particularly the numbers included in the zip codes. They compare their envelopes and those of the teacher to see if all the zip codes brought in are the same. They discuss the possible significance of the difference in numbers.

They examine the national directory of zip codes, looking up the codes of people they know in other cities. Using fake names and addresses in various cities throughout the United States, they address envelopes including zip codes.

Finally, they write letters to friends in other towns and use proper addresses (from home) and zip codes.

Learning about Fractions

19

General Overview Using fractions has become a fairly accepted numerical routine in our contemporary society, so this week's activities introduce students to the concept of fractions and develop skill in the use of fractions in everyday life.

Learning Objectives Students begin to use fractions.

MON

Objective Learning about fractions and what they mean: ½.

Teacher Preparation Provide several pieces of paper or cardboard of the same size, such as 3 × 5 or 5 × 7 cards. Have available objects that can be easily divided, such as oranges, apples, candy bars, and the like. For this lesson and most of the others on fractions, prepare flannel cutouts for use on the flannel board. Have one or more sets of disks to illustrate wholes, halves, fourths, thirds, sixths, and eighths.

Activities Students are introduced to the concept of fractions—something less than 1 or less than the whole thing. The teacher asks a series of questions as students perform certain activities: What is ½? We need half of an apple or candy bar. Students cut an apple or candy bar in half. Are both pieces the same size? How about half of this card (3 × 5)—students cut the card with scissors—are both sides the same size? Are they larger or smaller than the whole card? They cut halves and check the size. How many halves in the card?

Students use the felt cutouts in much the same way. They also show the 1 and ½'s on the fraction chart.

TUES

Objective Learning about fractions: ¼, ¾.

Teacher Preparation Have available the materials mentioned above plus a set of fraction flash cards that you can begin to use from time to time.

Activities The teacher introduces the fraction ¼ and then ¾. In much the same manner as in the previous lesson, students cut some objects first in half, then fourths. They show relative sizes and see that ¼ *is* smaller than ½, that it takes two ¼'s to make a ½ and four to make a 1. They use the felt board, the fraction chart, the chalkboard, and the cards.

Students then focus on ¾ and repeat many of the activities used above. For example, how big is it? Students observe that ¾ is smaller than 1 but larger than ½, that three ¼'s make ¾, etc.

WED

Objective Learning about fractions: ⅛, ⅜, ⅝, ⅞.

Teacher Preparation Have available all of the materials used in the preceding lessons for this week.

Activities Students are introduced to the ⅛ series by first cutting something into eighths. They compare and contrast the size to ¼, ½, ¾, and 1. They use the fraction chart to show that eight ⅛'s equal 1. They observe that ⅛ is smaller than ¼, because there are two ⅛'s in a ¼. They see the relationship of ½ to ⅜ to ⅝ to ⅞.

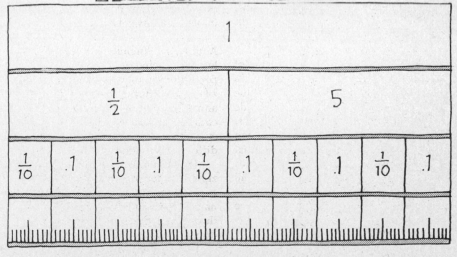

THURS

Objective Learning about fractions: ⅓ and ⅔.

Teacher Preparation You will need a new set of materials for this lesson because of introducing thirds. This additional fraction emphasis is considered necessary because of the frequency of use of the thirds in consumer affairs. Prepare a set of felt "thirds" similar to the "fourths" constructed previously, but limit the selection to disks (wholes), thirds, and sixths. Prepare also a fraction chart, this time having only three rows—1 (whole), ⅓, and ⅙. You will also need the materials from the first three fraction lessons.

Activities Students work in much the same manner as in the three previous lessons. They use a variety of objects made available by the teacher. They continue to practice until they exhibit an understanding of the relationships of ⅓ to 1, ⅓ to ⅔, and ⅔ to 1, and until they see the difference between and relative size of ⅓ to ¼, ⅓ to ½, ⅔ to ¾, and ⅓ to ⅛, etc.

FRI

Objective Learning about fractions: ⅙ and ⅚.

Teacher Preparation All of the materials from the preceding lessons of this week will be needed for this lesson.

Activities Students work with the fractions ⅙ and ⅚. Once again, they compare the differences and examine the relationships with regard to the other fractions studied and to 1.

Notes

20

MON

Objective Adding fractions with like denominators.

Teacher Preparation Have available the fraction felt cutouts and make a number line of fractions—one for ¼, one for ⅛, and one for ⅓.

Activities Students use the cutouts to begin to add fractions with like denominators and sums less than 1. For example, ¼ + ¼ = 2/4, 2/4 + ¼ = ¾, ⅛ + ⅛ = 2/8, 4/8 + ⅛ = 5/8, etc. Each problem is shown on the fraction line as well as in writing on the board.

TUES

Objective Adding fractions and mixed numbers.

Teacher Preparation Again, have the cutouts and fraction lines available for use in this lesson. Make a variety of flash cards to use as examples for students to work, such as 1½ + ½, 1¼ + ¼, 1¼ + 1¼, 2⅛ + ⅜, etc.

Activities With the felt cutouts and the fraction line, the teacher shows students how to add mixed numbers having like fractions. Students work examples using the cutouts and the fraction line. Once they can work examples using the aids, they use the flash cards for practice and work the problems on the board or on paper.

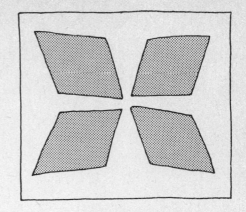

General Overview Students have been introduced to and are developing knowledge about fractions—what they mean and the relationships between fractions. This week's lessons teach them to use fractions—to add and subtract.

Learning Objectives Students learn to use fractions: addition and subtraction.

WED

Objective Adding fractions of all kinds.

Teacher Preparation Use the aids prepared for the preceding lessons and make additional flash cards for problems having fractions with sums greater than 1 and problems adding fractions with unlike denominators.

Activities The teacher shows students how to add fractions that result in numbers greater than 1. For example, $3/4 + 3/4$, $7/8 + 5/8$, $2/3 + 2/3$, $1 7/8 + 2 3/8$. Students then work examples, first with aids and then on the board. Finally, they use the flash cards for practice.

Next students are shown how to add fractions with unlike denominators: $1/2 + 1/4$, $1/3 + 1/2$, $1/8 + 1/6$, $1/8 + 3/4$, $2/3 + 7/8$. It is necessary to find a "common" denominator that may or may not be the product of the denominators. They use the aids and the flash cards for practice.

THURS

Objective Learning to subtract fractions with like denominators.

Teacher Preparation Use the materials provided for the preceding lessons and make some flash cards for examples and practice. Examples: $3/4 - 1/4$, $3/4 - 2/4$ $(1/2)$, $7/8 - 1/8$, $7/8 - 2/8$ $(1/4)$, $7/8 - 3/8$, $7/8 - 5/8$, $5/6 - 4/6$ $(2/3)$, $5/6 - 3/6$ $(1/2)$, $3/6 - 1/6$, etc.

Activities The teacher shows students how to subtract with like denominators using the aids to actually carry out the procedure: $3/4 - 1/4$, $3/4 - 2/4$, $7/8 - 3/8$, $5/6 - 2/6$, $5/6 - 4/6$, etc. Then, $1 1/2 - 1/2$, $4 3/4 - 1/4$, $7 1/4 - 3/4$. Students practice on other similar problems.

FRI

Objective Learning to subtract with unlike denominators.

Teacher Preparation Use the same aids but make more flash cards for examples such as: $3/4 - 2/3$, $5/6 - 1/4$, $1/2 - 1/3$, $1/3 - 1/8$, $7/8 - 5/6$, $1 3/4 - 1/3$, $2 5/8 - 3/4$, $6 3/8 - 5/8$.

Activities The teacher shows students how to subtract unlike denominators by finding a common denominator. Students then work examples until they gain proficiency in the techniques.

Notes

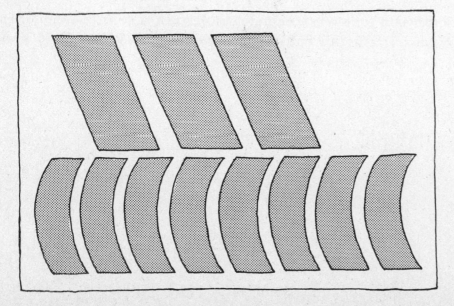

Understanding Percent

 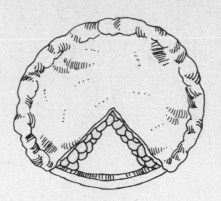

21

General Overview Many facets of daily life require the use of percent, from interest rates, to taxes, to service charges, to predicting the likelihood of rain. This unit delves into percent—what it means and how to solve problems using percent.

Learning Objectives Students understand and use percent.

MON

Objective Understanding the meaning of percent.

Teacher Preparation Make a "hundreds" board consisting of a square sheet of paper —poster paper, cardboard, or other— with 100 objects in rows of ten across and ten down. The objects can be disks, squares, triangles, bunnies, or anything else. Collect old newspapers and magazines for students to use in locating uses of percent. Make some charts to show how number facts can be written several ways: 8 out of 10 = $^8/_{10}$ = .8 = 80%; 6 out of 10 = $^6/_{10}$ = .6 = 60%; 25 out of 100 = $^{25}/_{100}$ = .25 = 25%; 50 out of 100 = $^{50}/_{100}$ = .5 = 50%; 75 out of 100 = $^{75}/_{100}$ = .75 = 75%.

Activities Students discuss the different ways to describe the same thing in numbers, using the charts prepared for that purpose. The teacher asks: What do you know about percent? Where have you seen it used? Let's find examples of percent being used in the newspapers and cut them out. The "hundreds" board is used to help students see how to figure percent. For example, 3% is 3 objects out of the 100; 30% is 30 out of the 100; 50% is 50 or ½ of the disks, and 75 is 75% of the 100 (and vice versa).

TUES

Objective Using percent in calculations of interest and service charges.

Teacher Preparation Locate information on interest rates and service charges of firms within your area—banks, finance companies, department stores, catalog-order firms, automobile agencies, and any others that have either of those charges. Bring in information from these firms, if available, and advertisements from newspapers and magazines.

Activities The teacher analyzes what it means to be charged 1.5% (1½%) per month interest or 18% per year. (For each $100 balance you pay $1.50 each month or $18 each year.) Other balances are used— $50, $60, $75. Advertisements: Pay 10% of balance per month—how much must you pay if you owe $150? Home loans— 9% per year—how much on a $15,000 loan? Service charge of ½%—how much on a $25 purchase? Pay interest of 1% per month—how much on a $50 account? Service charge of ¾% on purchase—how much on a $100 purchase?

WED

Objective Using percent in figuring discounts.

Teacher Preparation Collect advertisements that utilize discount selling. Also make a series of cards to represent store discount claims that students can use: 10% off on everything in the store; 15% off on coats; 25% off on suits; 5% off on all canned goods; oranges—33% off, etc.

Activities Students search the newspaper ads for "% off" notices and figure how much would be saved. They make bulletin boards showing the sales and how much could be saved by purchasing one or more of the items. They look into grocery ads, clothing-store ads, automobile ads, and others.

Using the "discount cards" for practice, students play games to see which team can get the most correct or get the correct answer first. Score one point for the team that "wins" in each problem, and then the teams figure the percentage each won of the total problems.

THURS

Objective Using percent in computing taxes.

Teacher Preparation Look up the taxes that must be paid in your vicinity: local sales tax, state sales tax, use tax, local property tax, local and state income tax, gasoline tax, and any others that may be collected. Make a chart showing the type of tax and percent due.

Activities The teacher explains each tax and students figure the amount that would be paid if: they earned $150 per week; they purchased $19 worth of groceries or a $35 coat (sales tax); they won $100; etc.

FRI

Objective   Using percent in communicating.

Teacher Preparation   Cut from newspapers, magazines, and any other printed source articles or notices that communicate in percentages such as: ''50% of the audience left early''; ''get 20% more mileage with 'mileagemore' ''; ''save 25% on your heating bill''; ''attend 80% of the time and earn a degree''; ''47% of the registered voters turned out''; ''the president received 51.8% of the vote.'' If articles cannot be found, make cards or charts to use with individuals or groups.

Activities   The teacher discusses the various percentage situations collected (or on cards) and students try to figure how many (and what) were involved (51.8% of the vote—how many?). Students look for ''percent communications'' in newspapers, magazines, books, and the like. They also communicate in their classroom with percent.

Notes

Learning Measurement: Weight

General Overview   One of the essential skills in measurement involves determining the weight of objects, a survival skill in this day and age of producer/consumer activity. This week's activities will involve the students in measuring by weight with all kinds of scales and balances, utilizing ounces, pounds, tons, and their metric equivalents.

Learning Objectives   Students learn to measure weight.

MON

Objective   Becoming familiar with the instruments used to measure weight.

Teacher Preparation   Locate as many different types of scales and balances as possible. Be sure to bring in a set of bathroom scales (several if feasible), kitchen scales, market produce scales, market meat scales, postal scales, balances from the high-school science rooms, upright school weight and height scales if possible, and any other types of scales that you can find. Also bring in old magazines, newspapers, and books that can be cut up. Have available drawing paper and pencils or crayons.

Activities   Students see and handle the various balances. They talk about each one: what you do with it, what it tells you, where it is used and why, how it works, and how you get information from each. The teacher explains the various dials and other indicators to read pounds and ounces and shows, where possible, that a pound has 16 ounces.

The students, with directions from the teacher, draw a pound and ounce ''time line'' with even spaces to show 16 ounces within each pound. The line is made long enough to include the weight of the heaviest student in the class.

Students search through old newspapers, magazines, and books to find and cut out pictures of scales and make posters from what they find.

TUES

Objective   Learning to use bathroom and upright medical-type scales.

Teacher Preparation   Have the various bathroom scales in the classroom and the school upright scales either in the classroom or available in the nurse's room or central office. Make a class weight chart with the name of each student printed on it and several spaces for weight entries across the chart.

Activities   Each student is weighed on the bathroom scales (several scales, if available) with other students reading the dial and recording the weight on a sheet of paper. They weigh to the nearest ounce and do the same with the upright scales. After all students have been weighed several times, their weights are recorded on the class weight chart.

WED   THURS   FRI

Objective Learning to weigh with kitchen scales.

Teacher Preparation Have several kitchen scales available, if possible. Have available several kinds of items that one might expect to weigh at home or at the grocery store, such as onions, apples, and potatoes. Have available poster paper and crayons and pens.

Activities Students weigh the articles available. Using various combinations of items, all students weigh them and record the weights on a chart. They compare weights and, if too many discrepancies occur, weigh the items again.

Objective Learning to use and/or read scales used in stores.

Teacher Preparation Make arrangements to visit a grocery store—preferably in small groups—to learn to read the produce and meat scales. Some hardware stores still weigh nails and feed/seed stores weigh their wares. If you can arrange to visit either of these, students can learn to use yet another scale.

Activities Students visit the stores as arranged. They weigh several objects (to the nearest ounce) that the groups bring to the store.

Objective Learning to weigh with all kinds of scales.

Teacher Preparation Have available all of the scales used for this unit. Have data-recording sheets for each student. Locate a variety of objects that can be weighed—nails, books, cans, balls, shoes, toys, the principal.

Activities Students weigh five to ten items, each time recording the item and its weight.

Notes

Learning Measurement: Distance

**23**

General Overview Another of the important applications of numbers which all children must encounter deals with the measure of distance. At this time, systems of measurement are becoming somewhat confusing because the change from English to metric is underway in the United States and some other countries. Measurement presented here is focused on the English system because its use dominates at this time, but metric is also included and should be used if the time is right.

Learning Objectives Students learn to measure distances.

MON

Objective Learning to measure distance using inches and feet.

Teacher Preparation Locate a variety of foot-long rulers—wood, plastic, different colors, and different markings. Also have available some six-inch rulers and rulers that are marked in both English and metric units. Buy, borrow, or make twelve-inch rulers for each student. Make or buy several cloth tape measures. Provide poster paper to make charts.

Activities The teacher talks with the students about measuring: "Everything can be measured. Some things are long, some short; some people are tall, some short. Handle the rulers; look at the scales and markings; point out inches and centimeters marked. What do you do with a ruler? Who knows?"

The teacher also leads students in some measuring activities: "Let's see who is the tallest in the room. Let's see how tall everyone is—and keep a chart! Each student, one or two at a time, stretches out on the floor and others measure their height. Write all measurements on the chart."

"How long is your foot? Let's measure and keep a chart. How long are your middle fingers? Your little fingers? Measure! How large is your waist—how far around? Use the tape measure to find out. Your upper leg? Your lower leg (calf)? Your forearm? Your upper arm? Your neck? Your head?"

TUES

Objective Learning to measure distance using the yard (and meter).

Teacher Preparation Have available enough yardsticks for each student and poster paper for charts.

Activities Students measure the length of the yardstick with their rulers or tape measures. Then they use their measuring devices to measure their classroom. Next they measure the football field and the gymnasium. They measure the same things using a meterstick or meter tape measure and they keep records and make charts whenever possible.

WED

Objective Learning to measure distances using the mile.

Teacher Preparation Have available a supply of road maps obtained from service stations, banks, the Chamber of Commerce, and similar groups. If possible, obtain (borrow) a pedometer for use for several weeks. Try to mark a mile somewhere close by the school—within sight, if possible, or block by block if that's the only way.

Activities Students participate in a discussion about large distances—from home to school, to the nearest towns, to the state capital, to the mountains, coast, lake, river, or desert. They go over the mile course marked off by the teacher. They walk the distance, if possible, but if not, they ride in an auto or on bicycle. They use the pedometer to see how far a mile really is.

Students use the road maps to figure out how far towns are from home; they "drive" 10 miles (16 km) down Highway 404 and show it on the map. They figure out how many miles it is to the coast or other significant recreation areas.

Students try to find out how far each lives from school; they make a chart showing the distance (miles and kilometers).

THURS

Objective   Using measurement to accomplish tasks.

Teacher Preparation   Have available the rulers, yardsticks, and all kinds of tape measures—cloth, plastic, and metal. Locate space for the measuring activities that seem suitable for your group.

Activities   Students mark off a garden—the teacher and students decide how large it should be and how far apart to mark the rows.

They use measurements in a variety of alternative tasks: measure out a Ping-Pong table; mark off the foundation for a house, a tent, a lean-to; mark off a basketball court, a baseball field, a swimming pool. Measure automobiles, bicycles, etc.

FRI

Objective   Using measurement of distance in games.

Teacher Preparation   This activity is a game in which students go certain distances and find a reward. Each location is different so each teacher must set up the game according to that situation.

Activities   Students play the game according to verbal or written instructions. For example, they start at the teacher's desk, go 10 yards toward the hallway, turn right and then 50 yards down the hall, turn left and go 40 feet to a tree, measure the distance around the tree, then 15 yards toward the parking lot and look under the bush! (The game could be set up within the classroom using smaller distances.)

Learning Measurement: Volume

General Overview In order to survive successfully in a contemporary society, one must be able to measure utilizing a variety of skills. Measuring volume is one of these. This week involves students in the use of the common volume-measuring devices such as the cup, pint, quart, half-gallon, and gallon.

Learning Objectives Students learn to measure volume.

MON

Objective Learning the relative sizes of the cup, pint, quart, half-gallon, and gallon.

Teacher Preparation Have available a set of measuring devices (preferably plastic) consisting of a cup, pint, quart, half-gallon, and gallon. Also have different-shaped containers in as many of those sizes as possible. Make labels with the name of each and glue on the container. Provide some pourable substance to go in the containers—water, sand, small gravel, or sawdust.

Activities The teacher has the set of measuring devices on a table and asks students what they are used for and what their names are. Which will hold the most? The least? Do you use them at home?

Students fill each container with water (if using material other than water, pour out piles to see the differences in sizes of the piles). They determine how many cups are in a pint and in a quart, how many pints are in a quart, how many quarts are in a half-gallon and a gallon, how many half-gallons are in a gallon, and how many pints are in a gallon. They make a chart or bulletin board displaying the various sizes of containers.

TUES

Objective Learning to use the cup and the pint.

Teacher Preparation Have available pint and cup containers. Bring in a variety of newspapers and magazines with advertisements and recipes.

Activities Students enact various situations requiring use of volume. Examples: (1) In a "school store," a customer needs a cup of sesame seeds, a cup of sunflower seeds, and a cup of aquarium sand. The customer also needs a pint of vinegar, a pint of pinto beans, a pint of milk, etc. (2) You are making a cake that needs a cup of flour, a cup of milk, ½ cup of powdered sugar, and a cup of brown sugar.

Students look through the papers and magazines and cut out ads and recipes that mention either the cup or pint. They make a collage or poster. They try to collect as many different shapes of pints as possible.

WED

Objective Learning to use the quart and the half-gallon.

Teacher Preparation Have available a a variety of quart and half-gallon containers. Bring in newspapers and magazines with advertisements that can be cut out. Have available poster paper for use in the student store plus felt-tip pens, tape, and the like.

Activities The teacher has a "quart" sale in the school store. Students decide which items can be sold by the quart and how much they should sell for and why. They look through the advertisements and locate as many ads as possible using the quart and half-gallon; they identify any good buys and determine whether a product is higher in the quart than the half-gallon size. Students then see how many different shapes of quart and half-gallon containers they can collect.

THURS

Objective Learning to use the gallon.

Teacher Preparation Have available gallon containers, newspaper advertisements, magazines, and old catalogs. Locate gallon capacities of various things such as water tanks, water heaters, and gasoline tanks in cars, motorcycles, boats, and airplanes.

Activities How is the gallon used? Students find as many uses as possible for the gallon container, looking in newspapers, magazines, and old catalogs. They discuss the uses and make bulletin boards showing what they find.

Students compare the gallon capacities of gasoline tanks of various types of cars, motorcycles, boats, and perhaps airplanes.

Students find out how many gallons of water in the city reservoir and other reservoirs of neighboring cities.

FRI

Objective Using volume measurement in purchasing.

Teacher Preparation Make poster advertisements of a variety of goods that can be purchased for normal consumption (use real cost figures so students can get accustomed to the larger container costing less per volume). Price items in the school store just as they would be priced in a real store—for example, with a pint costing 49¢ and a quart costing 94¢.

Activities Students "purchase" from ads and from the school store the best buy in terms of cost per volume. They use all sizes and prices—pint, quart, half-gallon, and gallon. Using advertisements from newspapers to study volume price differences (or visiting a real store, if possible), they look for price per volume. They compare, for example, the price of a pint, quart, half-gallon, and gallon of milk. (If you bought a gallon of milk in pints, how much would it cost per gallon? How about paint? Vegetable oil?)

Notes

Studying the Metric System

General Overview The shift from use of the English system of measurement to the metric system is now being encouraged by federal and state governments in the United States. We will convert totally to metric in the future and we must be able to make the transition with a minimum of confusion. This week's activities present an introduction to the metric system and opportunities to use it.

Learning Objectives Students learn about the metric system and how to use it.

MON

Objective Beginning to understand and use the metric system: millimeter, centimeter, and meter.

Teacher Preparation Collect several examples of current use of the metric system: road signs (speed and distance), weather forecasts (temperature and wind speed), analysis labels on foods (volume and weight), weights of anything, and descriptions involving metric language (accounts of a track meet, etc.). Prepare a chart showing the actual size of the units being studied this week: millimeter (mm), centimeter (cm; equivalent to 10 mm), meter (m; equivalent to 100 cm); milliliter (ml) or cubic centimeter (cc), and liter (l; equivalent to 1000 ml). Have available metric rulers of various lengths.

Activities Students participate in a discussion of the terms and symbols of metric measurement in general. They use the metric rulers to measure objects within the classroom, including the classroom, windows, and the like. They repeat many of the ideas and techniques found in week 23, but do so in metric language with metric tools.

TUES

Objective Learning to use the metric system: kilometer.

Teacher Preparation Mark off the distance of 1 kilometer—hopefully in an area that can be viewed visually but at least that can be walked (1 km equals approximately 3,281 feet). Make a chart showing the distance to neighboring towns using kilometers. Locate autos that have kilometer indicators on the dash (kph—kilometers per hour). Have available some road maps and kilometer scales to use in calculating distances. Introduce the term *kilometer* and discuss what it means.

Activities Students go outside to see and then walk the distance of one kilometer (.62 mile). Students bring in the distances to their homes and calculate it in kilometers. They look at a kph scale on an auto and discuss the meaning. They use the road maps to calculate distances to places they have been—the coast, mountains, grandparents, Uncle Joe's, etc.

WED

Objective Learning to use the metric system: volume.

Teacher Preparation Locate containers that have known metric volumes such as a liter, one-half liter (500 ml), and one-fourth liter (250 ml); try also to find a graduated cylinder (250–500 ml) that is marked off in milliliters. Bring in some cans and bottles from grocery stores that give the metric volume of the container. Talk about metric volume measurement—the liter and how big it is, the milliliter and how small it is, and how many milliliters are in a liter.

Activities Students look on the cans and bottles for the metric volume. They fill a number of containers to see how many milliliters or liters they hold; they use soft-drink bottles, milk bottles, and other containers that will be familiar. They use some bottles and cans that have the metric volume marked on the label. The teacher checks to see if they are correct.

THURS

Objective Learning to use the metric system: conversion to and from millimeter, centimeter, meter, and kilometer.

Teacher Preparation Be prepared to work with students in converting from English to metric and from metric to English. Make charts that give the conversion figures, both ways, and make comparison charts (actual size) so they can see the number of millimeters and centimeters in an inch and the number of inches in a meter. Have the metric and English rulers available plus a number of objects (straight edge) to measure.

Activities Students see the differences between the two systems of measurements and experiment with both. They convert from inches to milliliters and centimeters: "Close City" is 9 miles away—how many kilometers, etc.? They measure objects both ways and record the different figures. They discuss the conversion chart and convert from one system to the other and back.

FRI

Objective   Learning to use the metric system: conversions in volume.

Teacher Preparation   Have available the English volume-measurement instruments and the metric instruments. Make a chart to show the difference in actual size and make a conversion chart: liter to quart, quart to liter, pint to milliliter and vice versa, and liters to gallons and vice versa.

Activities   Students repeat many of the activities used above but use volume instead; they use the containers from Wednesday's lesson and measure both ways. They calculate gasoline tank size, gasoline purchases, and gasoline consumption using metric. They answer questions such as: At 60¢ per gallon, how much is gas per liter?

Notes

26

General Overview   This unit focuses on paying money for items and services. It provides real practice in paying cash, figuring the total amount of purchase, the amount of sales tax, and change due.

Learning Objectives   Students learn to pay for items purchased.

MON

Objective   Learning to use the penny, nickel, and dime.

Teacher Preparation   Students need practice in identifying and using money, so activities are suggested with this in mind. Make a large, attractive chart showing the penny, nickel, and dime with *relative* sizes simulating the real coins; use similar colors, too. Purchase play money or make pennies, nickels, and dimes for classroom use. Also have available a supply of real pennies, nickels, and dimes. Collect a variety of things for students to "buy" in the "store" costing from 1¢ to 24¢ and make flash cards with less than 25¢ change situations.

Activities   Discuss with students the three coins being studied so that they know the names, colors, sizes, and worth. With the play money students *count out* the amount for various situations that are presented either verbally or on flash cards. (For example, a piece of bubble gum, 2¢; a candy bar, 15¢ plus *tax*; a cookie, 7¢; two cookies, 13¢.) Students go to the store with a specified number of pennies, nickels, and dimes and purchase items available. Students role-play the salespeople or checkout clerks in order to gain experience with money.

TUES

Objective   Learning to use the dime and quarter.

Teacher Preparation   Make a chart showing the relative size of the dime and quarter plus what the quarter equals: 25 pennies = 5 nickels = 2 dimes and 5 pennies = 2 dimes and one nickel = 1 dime and 3 nickels, etc. Have available play money with the dime and quarter or construct them and make some flash cards with situations requiring the use of the dime and quarter. For example, a big cola, 25¢; a hotdog, 35¢; a knife, 75¢; a ball, 55¢; the number of pennies in a quarter, the number of nickels, etc.

Activities   As with the preceding lesson, the teacher discusses both coins with particular attention to the quarter and what it is equivalent to: 25 pennies = 5 nickels, etc. Students count out the dimes and quarters in play money to answer the flash-card questions. Finally, they purchase items in the school store, with either play or real money, as discussed in the preceding lesson.

Objective   Using coins and bills in purchasing.

Teacher Preparation   Collect all of the charts and all of the play money used in the first four lessons. Make a final set of flash cards to create real situations involving up to $50 or so for purchases of various kinds. Stock the school store with items costing up to $100 or so.

Activities   The teacher conducts a summary discussion with students, reviewing all of the money denominations shown on the charts. Using the flash cards, students count out the correct amount using any combination they want or specified combinations. For example, to pay $20 for a wagon: (1) they use any combination, and (2) they use only $1 and $5 bills. Finally, with the set of money, each student goes to the store to purchase a variety of items.

Notes

WED

Objective   Learning to use the quarter and half-dollar.

Teacher Preparation   Again, make a chart showing the quarter and half-dollar with illustrations of the value and composition of the half-dollar: 2 quarters = 5 dimes = 10 nickels and some of the other combinations. Make some "play" half-dollars or use some from play money and make flash cards with situations.

Activities   The teacher discusses with students the two coins, particularly the half-dollar and what it equals. Using the flash cards, students count out money (in quarters and half-dollars) to pay. Then, as above, they go to the store for practice in purchasing, using exact change of quarters and half-dollars.

THURS

Objective   Using $1, $5, and $10 bills.

Teacher Preparation   Make a chart to show each bill and its value; with the $1 bill show that it is equal to 4 quarters, 20 nickels, 2 half dollars, 10 dimes, etc. Show the $5 bill as five $1 bills and the $10 bill as ten $1 bills or a $5 bill and five $1 or two $5 bills. Also make (or purchase) some play bills in each denomination. Make a set of flash cards with money situations requiring the use of these three bills.

Activities   Conduct activities similar to those in the first three lessons. Using the charts, students discuss the bills, their value, where they come from, and the like. They use the flash cards for practice in using the bills. Finally, each student has a set of bills to use in purchasing items in the student store.

Purchasing Groceries

General Overview   One of the most necessary survival skills is purchasing groceries at a local store. Students need to know what to look for in terms of price per item, how to weigh items and utilize weight in purchasing, and how to *estimate* total purchases and then check the total bill. Most of the activities can be done in the ''class store,'' but a visit to a local grocery store would be worthwhile.

Learning Objectives   Students learn to shop at the grocery store.

MON

Objective   Learning to select the best buys in canned goods.

Teacher Preparation   Collect a variety of newspaper advertisements for grocery items from different stores and from several different weeks. Also have cans (full or empty) with prices intact.

Activities   Students use the newspaper ads of different stores (the same week) to select the best buys of several items: 1-lb. can of coffee vs. ½-lb. can vs. 2-lb. can; 14-oz. can vs. 28-oz. can. They publicize findings by making charts, tape-recording a broadcast, or writing a ''news release.'' Students then compare the same item for several different weeks. Do the prices go up, down, or up and down? They make a shopping list of canned goods (beans, corn, coffee, applesauce, etc.) and they find out how much the items cost per unit of weight—for example, how much for each ounce or pound?

TUES

Objective   Learning to select best buys in meats.

Teacher Preparation   Collect a variety of newspaper ads from grocery stores.

Activities   Students use the newspapers to carry on some comparative shopping. They determine the best buys in meat (price per pound). Each student makes a ''meat/cheese'' shopping list with quantities desired and finds the best buys from the ads. They determine how to get the most for their money.

WED

Objective   Practicing shopping for groceries.

Teacher Preparation   The school ''store'' should be ''stocked'' with grocery items: empty cans with labels and prices intact; meats of plastic (purchased) or made from papier mache; milk cartons; egg cartons; and plastic, papier mache, or real produce (apples, oranges, pears, potatoes, etc.).

Activities   Each student prepares a shopping list for three meals: breakfast, lunch, and dinner (the teacher needs to help plan these). They select the best buys from the class store or from a real store, figure the cost, check out at the cash register, and pay for the purchases.

THURS

Objective   Shopping for groceries.

Teacher Preparation   Contact a local grocery store to arrange a visit, taking small groups at a time. You do not need permission to visit the store, but it would probably be helpful to meet the manager and let him tell the group about the store. Prepare a shopping list with the group, including the grocery item, quantity, and size.

Activities   Students visit the designated store and do comparative shopping with the shopping list. (Hopefully they will be able to fill a shopping cart even though the items must be returned to the shelves.) They purchase the best buys based on cost/weight and figure the total cost of the goods.

FRI

Objective   Learning to estimate, total the grocery bill, and check out.

Teacher Preparation   Set up a variety of shopping lists for the school grocery store and duplicate; for each list set aside a fixed amount of play money. Try to have an adding machine (with tape) for use at the checkout counter.

Activities   Students shop for the items on their lists without exceeding the amount of money allotted. Students learn to keep a running tally of items selected, estimating how much is in the basket in terms of money. When the list is complete or the students think the money has been used, they total the list to see if they were right. If more goods can be purchased, they should do so; if they have selected too much, they should return some items. Then they actually check out and pay for the goods.

Notes

General Overview Effective, efficient purchasing requires understanding, skill, and ample practice in arithmetic. This week's activities create opportunities to practice arithmetic in the area of purchasing—clothing of all kinds and things related to the automobile such as gasoline, oil, tires, parts, and services.

Learning Objectives Students learn the arithmetic of purchasing: clothing and automotive accessories.

MON | TUES | WED

MON

Objective Learning the arithmetic of purchasing: clothing.

Teacher Preparation Collect newspaper advertisements for clothing for several weeks from several newspapers. Collect several mail-order catalogs. Stock the "student store" with possible clothing purchases such as shoes, socks, shirts, blouses, and the like. Have available blank sales slips or receipts for student use in selling and buying.

Activities Students participate in a discussion on what to look for in buying clothes—quality, reputation of the product and the store, price, guarantee, and the like. They discuss sales slips or receipts, including how to check entries, totals, and tax.

Individuals or groups decide what they need to buy today: two pairs of shoes; three pairs of socks; a shirt or blouse, a belt, etc. With this shopping list, students use two approaches: (1) they look through the newspapers and catalogs to select the items (again using factors such as quality, guarantee, reputation, price, etc.) and then fill out a sales slip, correctly making the proper entries, totaling, and adding any necessary sales tax; and (2) each student is allotted a specified amount of money to buy a specified number of items, again selecting from catalogs and newspapers and completing the sales slip. Some students "purchase" and some "sell," taking turns. They "pay" in cash or with some kind of credit.

TUES

Objective Using the arithmetic of purchasing: clothing.

Teacher Preparation The school store is the scene for activity for this lesson. Stock the store with items of clothing at varying prices. Have some of the same things at different prices to force students to make a choice (two pairs of shoes, for example). Have available the sales slips and play money for purchases and change and the forms required to make charge purchases.

Activities Students use the store for specified and unspecified purchasing: in some cases they have a shopping list; in others they "free" shop. Some of the time they have certain amounts of money to spend for certain kinds of items while they are unlimited for others. Students use the sales slips correctly, adding tax, and pay for the items as they would in any other store.

WED

Objective Learning the arithmetic of purchasing: automobile tires.

Teacher Preparation Collect newspaper ads and catalogs displaying automotive materials and supplies—gasoline, oil, tires, mufflers, batteries, services, tools, and others. Have sales slips and play money available.

Activities The teacher discusses with students the purchase of tires for automobiles: "What do you look for? How about price, mileage and other guarantees, reputation of tire and dealer, service by dealer, type (tubeless, tube type, white walls), balancing?" "Now you must purchase four tires, size C78-14. What kind should you buy? Look through the ads and catalogs—study the claims, analyze the prices, and buy four." "Now you have a limit of $100. Which tires will you buy? Which give you the most mileage for the money?" In all cases, students figure the correct price for the four, with sales tax and federal tax.

THURS

Objective Using the arithmetic of purchasing: other automotive needs.

Teacher Preparation Use the ads and the catalogs for Wednesday for this lesson and have available the sales slips and money. Duplicate a list of common automotive needs down the left side of the page and price columns across so that students can list prices from different sources for the various items. Include on the list batteries, oil, shock absorbers, mufflers, and others that might be pertinent for your car.

Activities After discussing the various automotive needs, students "shop" by looking into ads and catalogs for description and price and recording what they find. They notice that battery life varies and often the longer the life guarantee, the higher the price. Oil type and quality vary with price; muffler guarantees vary quite a bit, usually with the price. Students collect the information and then choose items to purchase. They total their purchases, adding necessary tax, and fill out the sales slip.

FRI

Objective Using the arithmetic of purchasing.

Teacher Preparation Select one clothing store and one automotive-supplies store to which you (or someone else) could take the class, preferably in small groups of five or six. Prepare a list of items that you have dealt with in class involving clothing and automotive supplies, and duplicate with enough for each student.

Activities Students visit the two stores and "shop" for the items on the shopping list. They get the various prices available and note the kinds available on all of the items on the list. Each student records the information on the sheet provided. Upon return to school, they discuss the shopping list, make decisions, and complete the totals.

Notes

29

MON

Objective Visiting at least one local bank.

Teacher Preparation Call one or two banks and make arrangements for a visit. If possible, arrange for someone to talk to the group for a few minutes and guide them on a tour of the facilities.

Activities Students visit the bank, paying particular attention to the way it is arranged for conducting business. They look for desks for customers to fill out forms, where the forms are kept, where customers go for transactions, where the tellers work, how they keep the money, the bank vault, and the like.

TUES

Objective Visiting the office of a savings-and-loan or finance company.

Teacher Preparation Call to make arrangements for a visit as mentioned in Monday's lesson.

Activities Visit the agency and look for the things mentioned in the preceding lesson.

General Overview   This is a preparation unit to get the group ready for a study of banking—checking accounts, savings accounts, and loans. Involved will be activities to motivate students as well as to prepare materials for use later.

Learning Objectives   Students study, construct, and furnish a banking agency.

WED

Objective   Designing a bank for the classroom.

Teacher Preparation   Locate a variety of drawings and photographs of bank customer-service areas. Have available paper and pencils for drawing plans for the classroom bank.

Activities   Students look at and talk about the picture collection and also talk about what they saw at the banks they visited. They draw diagrams of how they would like to set up the classroom bank. They design the counters, vault, etc., and discuss where each should be placed.

THURS

Objective   Constructing the furniture and equipment for the bank.

Teacher Preparation   You will need to locate and bring in a number of large, sturdy cardboard boxes or wood crates. Furniture companies, appliance stores, and other large-equipment firms are possible sources. Have something available for cutting the materials and, if you so desire, have tempera paint for the items constructed.

Activities   Individually or in small groups, students choose an area of the bank to complete and a furniture item to construct.

FRI

Objective   Supplying and using the class bank.

Teacher Preparation   Locate (construct or purchase) play money to be used in the bank. Bills of play money of various denominations are available for purchase and are advisable if possible because of the many occasions for use. If you do make you own money, bills should include $1, $5, $10, $20, and $50, and coins should include pennies, nickels, dimes, quarters, and half-dollars. You can mimeograph the bills using various colors of paper and cut the "coins" from different-colored posterboard. You will also need deposit slips, counter checks, and other forms used by the bank.

Activities   Students "stock" the bank and use it, making change for customers, counting money, going to the vault, using the burglar alarm, and the like.

Notes

Handling Checking Accounts

30

General Overview In order to succeed in our society, one must be able to handle money; one of the more acceptable techniques for doing so involves a checking account in a bank. Students should be able to open an account, deposit money, write checks, read and understand the monthly statements, and balance the checkbook with the statement.

Learning Objectives Students learn to work with and handle checking accounts.

MON

Objective Learning to open a checking account in a local bank.

Teacher Preparation Obtain from all local banks information on the checking-account options available through them. Borrow samples of the checks available and note charges for each. Make charts on each option for each bank. Show the minimum balance required, cost for each check, cost of printing the checks, and penalty for overdrawing.

Activities The teacher reviews the various checking-account plans available at different banks: the cost of each processed check, the cost of printing the checks, penalties for overdrawing, etc. Students decide to open an account at "bank X" (located in the classroom). With the play money, they open the account, giving name, address, social-security number (if known), the type of account desired, and the like. Each student deposits $50.

TUES

Objective Learning to deposit checks and cash into a checking account.

Teacher Preparation Locate deposit slips *and* checks for the various local banks included in Monday's lesson. Make copies, either Xerox or draw them on a stencil, for students to use for practice.

Activities The teacher provides the following directions:

1. You have earned $32.50 (play money provided by the teacher) from your part-time job and you need to deposit it in your checking account. First, fill out a deposit slip. Be sure the total at the bottom is the same as what you want deposited. Deposit it at your bank.

2. Now, you have two checks to deposit and you need $10 cash. How do you do it?

3. You have two checks and $18.50 in cash from another job. Fill out the deposit slip.

4. Now that you have deposited money in your account, how do you include it in your checkbook? You need to know at all times how much money you have so you will know how much you can spend. Total the amount in your checkbook. Practice this if additional skills are required.

WED

Objective Learning to write checks and maintain a correct balance.

Teacher Preparation Have available "real" checks from the banks being studied and also prepare some checks, larger than an actual check, on which students can practice. Make the practice checks about 4 by 9 inches and have a quantity available. Make up a set of money requirements—a bill to Suors & Rareback for $14.95; to U-Shock-'Em Electric Power Co. for $23.45; to Fix-It Hardware for $39.85; $27 for cash; etc.

Activities The teacher explains the need for money from the bank—in the form of cash or checks written to pay bills. Students use the money-requirement cards for "real" situations and write checks for each. They enter the payee *and* the amount of each check on the stub or accounting chart for the checkbook.

THURS

Objective Learning to keep a current balance tabulated in a checkbook.

Teacher Preparation Obtain from the local banks examples of the checkbooks they use—enough to have one for each student, if possible. Duplicate the stubs or record forms, in larger size than the original, so that each student can account for all the checks written.

Activities Using the checks (and sums) written for the preceding lesson and the beginning balance and other deposits, students keep account of the balance, check by check. They carefully account for each deposit (addition) and each check (subtraction) plus service charges, overdrafts, etc.

FRI

Objective Learning to balance a checkbook and check the bank statement.

Teacher Preparation Obtain "model" monthly bank statements to show to students. Mimeograph some statement forms and prepare a statement for each student showing checks written, deposits, and running balances. Be sure to show the beginning balance and ending balance.

Activities With the bank statement, each student looks at the checkbook to determine which checks have "cleared" the bank as shown on the statement. They check the balance after each transaction (checks written or deposit) and check the final balance.

Notes

General Overview Banking for most of us involves both checking accounts and savings accounts. This week deals with savings accounts—how to select one, how to open one, how to add to it, and how to keep up with the amount in it.

Learning Objectives Students learn to use a savings account.

MON

Objective Finding out what kinds of savings accounts are available.

Teacher Preparation Collect information on the savings-account plans available at banks in your area, including (1) percent interest, (2) when interest is computed, (3) withdrawal procedures and requirements, (4) how often the bank sends out statements, and (5) inducements to open accounts and make deposits.

Activities Using the information collected, students make charts to show differences between banks. Individuals or small groups take a different bank to "present" (or in some cases, the teacher could make the charts to present to them). After discussions of the various options, each student selects a bank for a savings account.

TUES

Objective Learning to open a savings account.

Teacher Preparation Obtain from local banks and savings-and-loan establishments the forms required for opening a savings account. If you cannot get enough forms to have one for each student, make your own.

Activities Each student completes the bank forms required to open a savings account. They make a small deposit to open the account in the classroom bank.

WED

Objective Learning to make deposits to a savings account.

Teacher Preparation Obtain deposit slips from the local banks and loan establishments. If you cannot get enough for use with your groups, Xerox or mimeograph additional copies.

Activities Using money that students "earn" within the classroom, they make a number of deposits, filling out the deposit slips correctly with name, address, account number, amount of deposit, etc.

THURS

Objective Learning to make withdrawals from a savings account.

Teacher Preparation Obtain enough withdrawal slips from various banks to have at least ten or twelve per student (you may need to mimeograph or Xerox your own).

Activities The teacher creates needs for money so that students will need to withdraw money from their savings accounts. With the help of the teacher, they determine the amount to withdraw, check to see if they have enough in their account, and fill out the withdrawal forms: name, address, account number, amount being withdrawn, etc.

FRI

Objective Learning to check a savings-account statement.

Teacher Preparation From your local bank files, locate a sample of the statements provided on savings accounts. Fill out one for each student to coincide with his or her account in the bank.

Activities The teacher shows the students how to check their statements: Are the withdrawal entries correct? Is the subtraction correct? Are the deposits correct? Is the addition correct? Was there any interest? If so, how much and was it added into your account correctly?

Notes

32

MON

Objective Becoming familiar with some of the opportunities for individual charge accounts.

Teacher Preparation Obtain information on a variety of charge accounts available throughout your area. Include descriptions of the accounts, what can be charged, credit limits, repayment schedules, interest and handling charges, and the like.

Activities Students make charts on the various plans available. They talk about how to use accounts, how much they cost, when the bills must be paid, and the like. Students then decide which charge plans they want to apply for.

TUES

Objective Applying for a charge card.

Teacher Preparation Obtain copies of applications for the variety of charge accounts discussed in the preceding lesson; Xerox enough applications to have at least one available per student (you can use oil companies, bank charge cards, store accounts, and any others available).

Activities Students fill out the applications, giving all the information requested. (In this type of role-playing situation, the teacher may need to make up social-security numbers, annual income, and the like.)

General Overview Another feature of contemporary America is the abundant opportunity for charge accounts or credit cards which most Americans utilize. Everyone should know about credit advantages and disadvantages and be able to open accounts if they so desire, make payments as required, and keep up with their expenditures, payments, and carrying charges.

Learning Objectives Students learn to select, open, and use charge accounts.

WED

Objective Charging one item with a credit card.

Teacher Preparation Your classroom store and other commercial ventures must come into play here and in a variety of occasions later. Perhaps the classroom bank can be modified! Make cardboard credit cards to be issued (upon receipt of the application) to each student and have available the company sales-receipt forms.

Activities Students purchase goods from the appropriate business such as gasoline, food, or other items. The "salesperson" fills out the company sales receipt and the "customer" checks it for mathematical accuracy and then signs.

THURS

Objective Charging several items with credit cards.

Teacher Preparation The same as the preceding lesson.

Activities Students purchase a variety of items from one store, and then perhaps they use another card at another type of establishment.

FRI

Objective Learning to check charge-account statements.

Teacher Preparation Using the charge records for purchases made on Wednesday and Thursday, make "monthly statements" to be "mailed" to each student.

Activities Students check their bills using their receipts from the various purchases. They check each entry, the addition, handling and interest charges, dates, and the like.

Notes

Making Loans

33

General Overview Contemporary America has adopted a way of life that has the borrowing of money as an essential element. Along with this element have developed lending agencies ranging from banks to savings-and-loan establishments to finance companies of various kinds. Students must be able to select the best lending agency to apply to for loans and to handle repayment.

Learning Objectives Students learn to make loans.

MON

Objective Learning how to select the best lending agency.

Teacher Preparation From your visits to local banks, pull together the information on loans from those banks and visit (or call) finance companies and savings-and-loan firms to get information on (1) availability of money, (2) procedures for applying for money, and (3) interest rates. Make charts showing this information for each agency plus charts to highlight the differences between agencies.

Activities Students look at the different charts and discuss various features about the lending agencies—how much it "costs" for the money, when they must pay it back, kinds of loans, etc. They try to decide on the best establishment to apply for a loan.

TUES

Objective Applying for a small loan from a bank.

Teacher Preparation From your banking file, pull the forms required to apply for small loans from a local bank. Have enough copies so that each student can fill one out.

Activities Students discuss their "needs," pehaps to buy a car, go to Europe, or buy a television set. Together they figure out how much it will cost to borrow or use the bank's money for a year ($1,000 will cost $100 each year at 10 percent interest). Next, they apply for a loan, have it approved by the bank, and get the money from the bank.

WED

Objective Applying for a small loan from a finance company.

Teacher Preparation This activity is almost like the preceding lesson, with the purpose of showing the difference in *cost* of the money from different types of lending agencies. Most finance companies must charge high interest rates—frequently higher than banks—because they tend to take more risks in lending to more "high-risk" individuals. Have finance-company forms available for each student.

Activities Students perform the same activities as in the preceding lesson, but they observe the difference in the total cost of using finance-company money because of the higher interest rates.

THURS

Objective Making payments on loans.

Teacher Preparation When students receive the loan, set up a rate of payment and make sure they realize that it must be paid back. Also set up a system of accounting—a bankbook, a chart, or any other technique used by the various agencies. Have available an "account" for each student.

Activities Students write a check or pay cash for their loan payment and enter it on the proper ledger. They keep a running account of how much has been paid.

FRI

Objective Learning about applying for large loans.

Teacher Preparation Students should know that people can borrow from lending agencies for large amounts of money over long periods of time. Have available pictures of houses for sale—real-estate ads perhaps—with prices; also have information on home loans available locally —amounts available, interest rates, and the amount of money for a down payment.

Activities   The teacher talks with students about how much money it takes to buy a house, using the real-estate clippings as a focal point. They decide which one to "buy" and then figure out how much the down payment would be, how much interest would be involved, the size of the monthly payments, and the like. (Call a realtor if you need help in finding or figuring out some of the information needed.)

Notes

General Overview Weather is one of the factors of our environment that is ever present but also ever changing. It affects us in thousands of ways and it *is* important. We cannot all learn to predict the weather accurately, but we should be able to handle the information that is available and make decisions concerning our own behavior to fit the data. Thus this unit deals with some basic weather data.

Learning Objectives Students learn to observe weather conditions.

MON

Objective   Learning to read and use liquid-type thermometers.

Teacher Preparation Locate a variety of liquid-type thermometers, either alcohol (usually red liquid) or mercury (silver colored) that you can keep in the classroom for several weeks. Make a "ribbon" or "zipper" thermometer to use with individual students in reading thermometers. For example, draw a thermometer scale (Celsius and Fahrenheit) on cardboard and attach a red ribbon or a zipper that can be moved up and down to represent the red liquid.

Activities Students observe the various thermometers and take them to various locations to see what happens: in direct sunlight, under a cool tree, in a closet, in a very warm room, in the furnace room, etc. They talk about the changes and what they mean. They discuss what makes the thermometer work and why—how it is made, what is inside, etc. Students use various temperatures (both C and F) on the demonstration thermometers (ribbon and/or zipper): 0°F, 32°F, 212°F, and the present inside temperature and outside temperature. They practice until they can read the temperature quickly and correctly. Students keep a record of the morning and afternoon temperatures in the classroom, using a poster chart to record the data.

TUES

Objective Learning to use and read oral thermometers.

Teacher Preparation   Obtain from the school nurse (or elsewhere) several oral thermometers. Have available poster paper to record temperatures.

Activities Each student takes and reads his or her own temperature (several times so that they can see the consistency of human temperature). They make a chart showing the various temperature readings for each student in the class. They take readings at least once a day for approximately ten days and they discuss the thermometer readings and the significance of the "fixed" temperature to humans.

WED

Objective Learning to read and interpret various kinds of mechanical thermometers.

Teacher Preparation Locate meat thermometers, oven thermometers, freezer thermometers, and old thermostats, if possible. Find the temperature indicator in an automobile that will be available to the group. See if the custodian can show the group thermometers and gauges associated with the heating and cooling system in the school, on the furnace itself, and the thermostats in the various rooms. Locate manuals that tell how to cook meats with meat thermometers, temperatures for oven use, freezer temperatures, and the like.

Activities Students examine, study, and discuss the various mechanical thermometers in the classroom: how they work, what they tell you, where they are used.

Using the manuals, students determine the thermometer reading for various meats such as a 10-lb. roast, a 12-lb. ham, and a 15-lb. turkey. They show on the thermometer where the indicator would stop for the various meats and/or draw pictures of the face of the thermometer to show the indicator needle. They do the same for other foods baked in the oven and show where the needle would rest.

They visit with the custodian to see and discuss the room thermostats and gauges on air conditioners and furnace. They also examine temperature indicators on cars.

THURS

Objective Learning to interpret weather reports.

Teacher Preparation Collect information on weather stations, the U.S. Weather Bureau, and any other facilities and agencies that collect data on weather conditions. If there are local resource people who are experts on meteorology, see if they will come to visit your classroom or let the class visit them. Bring the weather report from the local newspaper for several weeks.

Activities Students participate in a discussion on how weather information is collected and what kind of information is used: barometric pressure—inches of mercury (30.10, for example); wind speed—13 mph from the north-northeast; rainfall—2 inches (in the rain gauge); relative humidity—80%; etc. They visit a weather station or let a local expert visit the classroom to talk about reading various instruments. Students keep a record of the weather "number" data. They make a chart for temperature, wind speed, barometric pressure, and relative humidity.

FRI

Objective Making weather instruments and recording the data.

Teacher Preparation Collect materials for classroom construction of wind-speed indicators, wind-direction indicators, bottle barometers, and rain gauges. Locate, in one of any number of elementary science reference books, instructions for making these instruments (also see week 21 in the Science section of this book).

Activities Students make the instruments suggested above and use them, recording the information on charts and in graphs.

35 MON

Objective Learning to compute miles per gallon of gasoline.

Teacher Preparation Collect information on miles of travel that can be expected and/or actually occurs from a gallon of gasoline. Expected claims for "mpg" can be obtained from car dealers; actual mpg can be obtained from anyone who drives a car and keeps records on gasoline consumption. Gather information on a variety of types of automobiles (Ford, Chevrolet, Buick, etc.), a variety of sizes, and a variety of types of vehicles (cars, trucks, motorcycles, etc.). Get information on "in-town" and "on-the-highway" driving.

Activities Students examine and discuss the "mpg" information collected by the teacher and also try to collect from relatives and friends mpg data on their automobiles. They make comparison charts and determine which cars would be the most economical in terms of gasoline usage. (Some students in the class figure "kpl"—kilometers per liter—and compare these figures to the mpg figures.)

General Overview Arithmetic is essential for survival in many facets of contemporary society, but in a highly mobile environment no aspect is more significant than travel. Everyone must be able to travel, and travel involves arithmetic whether a person drives a personal vehicle or utilizes some form of mass transit. This unit examines some of the arithmetic of travel and provides opportunities for students to practice with travel arithmetic.

Learning Objectives Students understand and use the arithmetic necessary to travel.

TUES

Objective Figuring the cost of gasoline fuel for automobiles.

Teacher Preparation Keep the information from the preceding lesson and make a chart showing the price of gasoline per gallon and per liter at local gasoline stations.

Activities With the price information provided by the teacher and also that collected by the students, each student figures the cost of gasoline per mile for the various brands. (Example: the Ford LTD gets 15 miles per gallon; Exxon gas costs 65.9¢ per gallon; 65.9¢/15 = 4.4¢ per mile.) They make charts showing comparative costs with various brands (and prices of gasoline). (Some students figure the cost of gasoline per kilometer using the same techniques.)

WED

Objective Learning to compute expenses for an automobile trip.

Teacher Preparation Have available the information collected in the two preceding lessons plus a number of road maps with mileage charts and mileage scales.

Activities Students determine several destinations for "make-believe" trips and look up or compute the distance to those destinations. They then figure how much the gasoline will cost for the various trips with various cars, trucks, and motorcycles.

THURS

Objective Learning to compare the cost of automotive travel to other means of travel.

Teacher Preparation Locate cost-per-mile figures to operate the "average" American automobile. Automobile associations maintain these figures and they include *all* costs in these figures—gasoline, oil, depreciation, maintenance, tires, and the like. Many businesses with mileage expenses (salesmen on the road) will also have this information. Also locate cost of travel by airplane, train, and bus—per mile and by destination.

Activities Students make charts to show and compare costs of travel by the different modes. They compare by mile and also by trip. They consider *all* costs for each trip, including travel to the airport from home and to the destination, parking fees, meals while traveling, and motels if necessary.

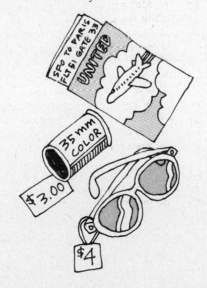

FRI

Objective Identifying aspects of travel requiring a knowledge of and skill in arithmetic.

Teacher Preparation Locate information on any aspect of travel in your region that requires arithmetic—for example, toll charges, bus tickets, taxi fares, airline tickets, tram tickets, and subway tickets. Try to find out the amount for the various fees and get old tickets if possible. Locate pictures or diagrams of highway speed-limit signs; note the "number" signs in use in your vicinity. Have available a variety of road maps covering your state and those states immediately surrounding your state.

Activities Students discuss all of the "travel" arithmetic they can remember and, with help from the teacher, they make lists of all of these; each student then selects one or more to study in depth. They report back to the group and discuss their findings.

Notes

Using Arithmetic Games

General Overview One technique often highly effective with all types of groups is gaming. Although playing games will not solve all of the problems of classroom management, games are excellent for change of pace and are particularly effective in providing practice for skill development. The games described in the next five lessons (days) represent one type of game patterned after popular sports games. There are other types of games—board games, games of logic, and many more—but the emphasis here is on games to assist in skill development with a strategy easy to understand and use.

Learning Objectives In each case (with each game), students will attempt to increase skill in arithmetic fundamentals through meaningful practice.

MON

Objective Increasing skill development through the use of "arithmetic football."

Teacher Preparation The game of "arithmetic football" is played by two teams with any number of players on each team, from two to twenty-five. Mark on the chalkboard (or have a poster) the football field. Use a scale such as ¼ inch to a yard to represent a field 100 yards long (or the metric equivalent) with 10-yard intervals (or the equivalent so that there are ten equal sections down the length of the field). Develop a quantity of arithmetic problem cards for use in the game, the type depending upon the skill area being studied (i.e., multiplication, addition, subtraction, or division). For all problems, assign a "gain" or "loss" value —for example, the problem card "5 + 3" might represent 3 yards gained if answered correctly or 3 yards lost if incorrectly answered, and the card "25 + 36" might represent 10 yards (first down). If the answer is a correct one, the team advances the assigned yards; if incorrect, they *lose* the yards. Have separate decks with varying levels of difficulty so that students may choose the level of difficulty by the number of yards to be gained (or lost if answer incorrect). For variation, have a deck of "punt" cards and a deck of "pass" cards, again each with assigned yard values.

Activities Divide the class into two teams, each with a name, and toss a coin to see which team begins. Students play as described above for a specified period of time. The team ahead when time runs out is the winner.

TUES

Objective Increasing skill development through the use of "arithmetic baseball."

Teacher Preparation "Arithmetic baseball" may be played with any number of players divided into two teams. Mark off a baseball diamond on the chalkboard (or have a poster already prepared) and label home plate, first base, second base, third base, and pitcher's box. Develop a large number (100 to 200) of problem cards for use in the game, the type of problem depending upon the skill area and level of difficulty being stressed. Divide the cards into four piles, the easiest problems in the pile to be used for a "single," the next for a "double," then "triple," and the most difficult for a home run.

Activities To play "arithmetic baseball," divide the class into teams and let each team select a captain and establish a "batting" order. Play for nine innings (times "at bat") or for a specified time established before play begins. If the answer to the problem is correct, that student gets to advance on the diamond: one base for a "single," two for a "double," etc.

To make the game more complicated, have each "hit" (correct answer) challenged by attempts at being "caught" or "thrown out." Each time a problem is answered correctly for a "hit," a student on the other team gets to try to answer another problem and, if correct, gets the first player "out." To play this variation, have two problems on each card, one for the "batter" and one for the "fielder" (catcher).

WED

Objective Increasing skill development through the use of "arithmetic basketball."

Teacher Preparation The game of "arithmetic basketball" is played with two teams and any number of players. Mark off a basketball court on the chalkboard or on a poster—draw in the two baskets. Develop a series of problem cards to provide practice in the skill area that needs work. Divide the cards into three stacks with one stack the easiest and the third the most difficult. The easiest problems— "lay-ups"— can be blocked by either of two opposing players answering the "block" problems. Each lay-up card has a problem to be answered for 2 points plus two problems (slightly more difficult) that may "block" the lay-up. Either may be answered correctly to block. For example, player on the "Celtics" answers a lay-up problem correctly. Now two players from the "Nuggets" must answer one of the other two problems on the card to "block" the lay-up. If one or both answer correctly, no points are awarded.

For the next most difficult stack (a 10-foot jump shot), there is only one opportunity to block (with a block question, slightly more difficult on the card). For the most difficult problems, there can be no blocked shots. If the problem is answered correctly, the team automatically gets 2 points.

Activities Divide the class into two teams and let the teams select a captain and decide the order of shooting. The captain and players decide which of the "shots" to try each time. Play rotates from team to team whether there is a score or not.

THURS

FRI

Objective Increasing skill development through "car" and "motorcycle racing."

Teacher Preparation The teacher and class must decide whether to race cars or motorcycles and where to race. A car race from the east coast of the United States to the west coast is an exciting race. On a large map of the United States, select a course across the nation. Depending upon the class, decide how difficult to make the course. Possibilities: (1) the course could "weave" and "wind" all over the nation, going through many states; (2) the course could go straight across—from New York to San Francisco; (3) the course (either 1 or 2) could be marked off in 100-mile intervals. After establishing the course, develop sets of problem cards to provide practice in skill areas needing emphasis. Have at least three levels of difficulty: the easiest to represent one state along the route or 50 miles; the next for two states or 100 miles, and the most difficult for three states or 200 miles.

Activities Divide the class into teams of three or four each and let each team represent one car (or motorcycle). They may want to draw or cut out pictures of their vehicles and mount them on cardboard to move across the map. With a correct answer to the problem cards, they move ahead one, two, or three states (or 50, 100, or 200 miles). If incorrect, they move back the specified amount! Teams take turns in answering questions. The winner is the first team to reach the destination.

Objective Increasing skill development through the game of "sailboat racing."

Teacher Preparation Select a lake or ocean on which to hold the "race." With some groups a "round the world" race would be fun and would also help them to learn more about the geography of the world. Any area could be used, though, and students could learn about the area as well as improve arithmetic skills.

Develop problem cards to focus on the skill area being studied. Have at least **three** different levels of difficulty for three decks of cards. Depending upon the length of the race course, assign mileage (kilometers, if metrics being used) to each deck: if an ocean race around the world, for example, each correct answer for the easiest problems could be worth 500 miles; the next 1,000, and the most difficult 1,500. After mapping the course and preparing the cards, you are ready to begin the race.

Activities Help youngsters divide into teams or "crews" for the various vessels in the race. They should name each vessel, draw or cut out a picture of her, and get ready to race! Each crew should have a "skipper," and the skipper and crew decide which deck of cards to attempt at each turn.

Notes

Notes

Notes

Notes

Notes

Notes